Ecosystems & Biodiversity

David Holmes

Series Editor: Sue Warn

Philip Allan Updates
Market Place
Deddington
Oxfordshire
OX15 0SE

Orders
Bookpoint Ltd, 130 Milton Park, Abingdon, Oxfordshire OX14 4SB
tel: 01235 827720
fax: 01235 400454
e-mail: uk.orders@bookpoint.co.uk

Lines are open 9.00 a.m.–5.00 p.m., Monday to Saturday, with a 24-hour message
answering service. You can also order through the Philip Allan Updates website:
www.philipallan.co.uk

Front cover photograph reproduced by permission of Corel

Printed in Great Britain by CPI Bath

Philip Allan Updates' policy is to use papers that are natural, renewable and
recyclable products and made from wood grown in sustainable forests. The logging
and manufacturing processes are expected to conform to the environmental
regulations of the country of origin.

Contents

Part 4: Coastal ecosystems

Part 5: Freshwater ecosystems

Part 6: Ecosystems of low primary productivity

Part 7: Human ecosystems

Contemporary Case Studies

Part 8: Biodiversity: loss and conservation

Part 9: Examination advice

Index

Introduction

About this book

This book begins by highlighting the importance of the world's ecosystems, which, collectively called the **biosphere**, contribute a range of goods and provide extremely valuable services towards the survival of humankind.

The means of ecosystem classification is discussed using the concepts of scale (i.e. from a single tree or rock to a global biome) and also the degree of human influence — from natural forests or grasslands through to completely managed agro-ecosystems. Levels of primary productivity are also considered. These have a direct impact on the complexity of food webs and food chains, as well as the variety of ecological niches.

Also examined are the issues of ecosystems under threat, both directly (e.g. by deforestation) and indirectly through processes such as global warming or the introduction of alien species. In many cases, there has been a marked loss of biodiversity, through either the complete destruction or the modification of ecosystems. Other threats arise from the rising world population, exploitation by transnational corporations and the need of many of the poorest people in the world to feed themselves. Inevitably, the destruction of goods such as trees, or animals for meat, means that valuable services cannot be provided — hence the need for sustainable management strategies. Examples of such strategies are included in the final part.

Understanding ecosystems requires a sound grasp of their structure and functioning, i.e. how they work and change over time through the process of **succession**. Key terms, which are emboldened in the text, are explained on pages vii–viii.

The main part of the book describes a wide range of global ecosystems and biomes, broadly classified according to the WRI (World Resources Institute) system. A standard format is used for each ecosystem group, focusing on its relative importance, distribution, threats and management. Table 1 (page 5) summarises the selection of ecosystems and lists the main case studies.

Part 8 explores the global issue of biodiversity loss, assessing the scale of the problem and evaluating attempts to conserve and protect important habitats. Various strategies are discussed, including 'hot spots', eco-regions and sustainable management, using a range of case study material.

The final part of the book provides examination advice, including standard eco-systems questions (both long and short essays) and guidance on handling resources.

Key terms

As a basis for any examination preparation, a good working knowledge of the key concepts and terms is essential.

Ecosystems

Biomass: the total amount of organic matter — the higher the trophic level in a food chain the less the biomass (biomass is usually expressed as g/ha).

Biome: a major terrestrial ecosystem of the world also known as a province, biochore or region.

Biotic and abiotic: the living and non-living components of an ecosystem respectively.

Climax community/vegetation: vegetation that is representative of the final stage of succession — this vegetation is in equilibrium with the local or regional climate.

Community: an assemblage (or grouping) of particular plants and animals that are linked by the flow of energy, the cycling of nutrients and the regulation of population within an environment.

Ecocline: a type of landscape boundary, but also a gradual and continuous change in the environmental conditions of an ecosystem or community. It contains communities that are considered more environmentally stable than those of **ecotones**.

Ecosystem: a system of which both the living organisms and their environment form components (elements) — these components are linked together by flows and are separated from the outside by a boundary.

Ecotone: a transition area between two adjacent ecological communities (ecosystems). It may appear on the ground as a gradual blending of the two communities across a broad area, or it may manifest itself as a sharp boundary line.

Goods and services: 'Goods' are direct products that can be derived from an ecosystem; 'services' are the benefits that the ecosystem provides.

Habitat: the place where a particular species lives and grows. It is essentially the environment — at least the physical environment — that surrounds, influences and is utilised by a particular species.

NPP: Net primary productivity — the difference between the rate of conversion of solar energy into biomass in an ecosystem and the rate at which energy is used to maintain the producers of the system.

Pioneer species: the first plant species to colonise previously uncolonised land. Since uncolonised land usually has thin, poor-quality soils with few nutrients, pioneer species are typically very hardy plants. Pioneer species may include lichens, which grow on bare rock surfaces (lithosere **succession**).

Plagio-climax: where the process of succession has been stopped or deflected by human activity.

Primary consumer: an organism that feeds directly on producers, and so is normally found in trophic level 2 of the pyramid.

Succession: the gradual and predictable change in plant and animal species over time, for example bare ground is colonised by plants and there is a series of sequential replacements as one set of dominant plants replaces the other.

Trophic level: a position in the food chain, determined by the number of energy-transfer steps to that level.

Biodiversity

CITES: the Convention on International Trade in Endangered Species of Fauna and Flora is an international agreement between governments. Its aim is to ensure that international trade in specimens of wild animals and plants does not threaten their survival.

Diversity: usually taken to mean a measure of the number of different types of species in a given area.

Endemic species: exclusively native to a particular place or region. Endemic species tend to have a high conservation value.

Keystone species: a species whose very presence contributes to a diversity of life and whose extinction would consequently lead to the extinction of other species.

LPI: the **Living Planet Index**, created by WWF, which displays data derived from trends over the past 30 years in the population of hundreds of species of birds, mammals, reptiles, amphibians and fish.

MPA: Marine Protected Areas (MPAs) are areas designated to protect marine ecosystems, processes, habitats, and species that can contribute to the restoration and replenishment of resources for social, economic, and cultural enrichment.

RAMSAR: the Convention on Wetlands, signed in Ramsar, Iran, in 1971, is an inter-governmental treaty that provides the framework for national action and inter-national cooperation in the conservation and wise use of wetlands and their resources.

An introduction to ecosystems

About ecosystems

Defining the ecosystem

While there are many definitions of **ecosystems**, all of them have certain components — organisms, sun, land, air and water (Figure 1). More formally, we can think of an ecosystem as 'a complex set of relationships among the living resources, habitats and residents of an area, including plants, birds, trees, fish, microorganisms, soils, water and people'.

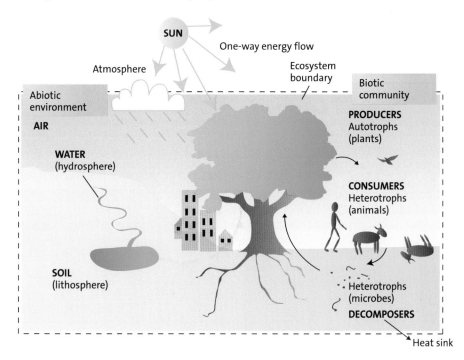

Figure 1
A functional diagram of an ecosystem

The value of ecosystems and their goods and services

Ecosystem services are benefits that people obtain from ecosystems. These include provisioning, regulating and cultural services — which affect people directly — as well as supporting services needed to maintain the other areas. Changes in these services can affect human well-being through impacts on security, the basic materials for a good life, health, and social and cultural relations. These constituents of well-being are, in turn, influenced by, and have an influence on, the freedoms and choices available to people.

Figure 2 Basic goods and vital services provided by ecosystems

BASIC **GOODS** FOR SURVIVAL	VITAL **SERVICES** FOR SURVIVAL
Food crops, e.g. tree crops such as fruit and nuts	Climate regeneration
Food via agroecosystems and also indirectly for feeding livestock (fodder)	Air purification: trees remove CO_2 (carbon sinks) and emit O_2 (green lungs)
	Water control: vegetation exerts an impact on the water cycle
Meat and fish	Flood protection (forested watersheds) and storm protection (mangroves and reefs)
Building materials, timber, paper etc.	Water purification: dilutes and carries away waste
	Cycling of nutrients
	Generation of humus to produce soil
Source of water	Maintenance of biodiversity
	Provision of wildlife habitats
Energy via biomass and HEP	Opportunities for aesthetic enjoyment and recreation
	A gene pool for the future
Genetic resources for medicine etc.	Employment opportunities across all sectors

ECOSYSTEMS

Ecosystem goods are products that can be derived directly from the ecosystem. These might include timber from trees, or meat and fish for eating. Goods are generally easier to value (in a monetary sense) than services. While some goods can be sustainable, the use of others (e.g. hardwood timber) destroys the realisable role of services. Figure 2 shows the basic goods and vital services related to ecosystems.

It is obvious that the world's global economy and national economies are based on the goods and services derived from ecosystems. It is also evident that human life depends on the continuing capacity of the biosphere's ecosystems to provide their multitude of benefits. In reality, there is a complex wheel of benefits derived from ecosystems — a mixture of goods and services (Figure 3).

Figure 3 Benefits of ecosystems

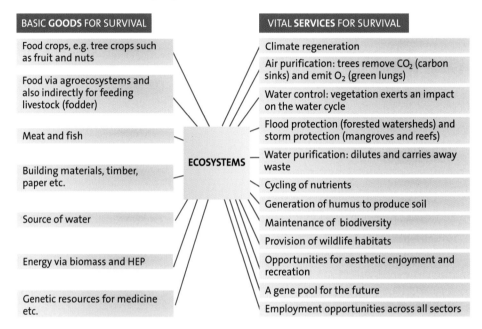

Ecosystems as a system

The biosphere is the global ecosystem. Within the biosphere there are **inputs**, **outputs** and areas for **storage** or processing. Relationships through which the outputs can have an effect on the inputs are called **feedback loops**.

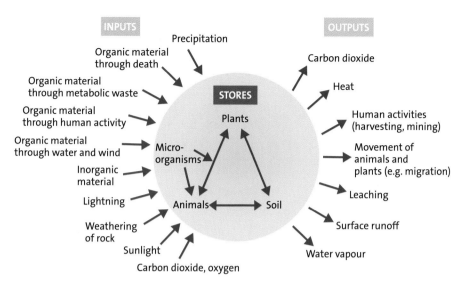

Figure 4 Ecosystems as 'systems', with inputs, outputs and stores

The 'systems' approach can be used to understand ecosystems at a range of scales (Figure 4). You do need to be selective, however, as some of the inputs and outputs shown in Figure 4 will not be appropriate for the particular ecosystem you are studying.

All natural ecosystems are open systems, where inputs and outputs can be freely exchanged with other ecosystems. Ecologists have identified a number of ecosystems at different scales, as shown in Figure 5.

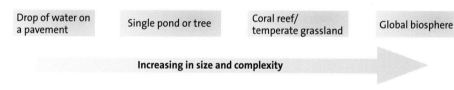

Figure 5
The differing scales of ecosystems

Global ecosystem distribution

Figure 6 shows the distribution of the world's major ecosystem types (or **biomes**). Note that you will need to refer to this map on many occasions throughout this book.

The abundance and distribution of biomes is principally controlled by climate — in particular, average temperature and the availability of moisture (Figure 7).
- Forest vegetation is only found in areas of high moisture availability.
- Grassland (including savanna) and scrub can be found in areas of lower moisture availability.
- Tropical rainforest, savanna and scrub are restricted to areas with warm climates.
- Coniferous forest and tundra vegetation are only found in regions with cooler climates.

'Limiting factors' are important in controlling the distribution of these vegetation types.

The exact boundaries of ecosystems are often difficult to establish, as one biome tends to merge gradually into another. The rate of boundary change (or **environmental gradient**) is greater in some cases than in others. Obviously, a freshwater pond has a clearly defined physical and ecological boundary. In contrast, the margin of the area

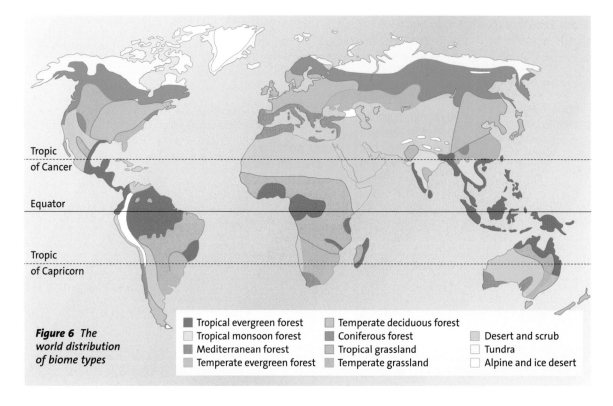

Figure 6 *The world distribution of biome types*

Legend:
- Tropical evergreen forest
- Tropical monsoon forest
- Mediterranean forest
- Temperate evergreen forest
- Temperate deciduous forest
- Coniferous forest
- Tropical grassland
- Temperate grassland
- Desert and scrub
- Tundra
- Alpine and ice desert

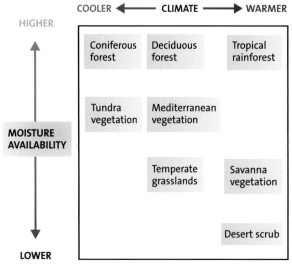

Figure 7 *Factors controlling the distribution of biomes*

where particular birds feed is much harder to distinguish. These 'blurred' edges are commonly called **ecotones**.

At regional and local scales, human action and physical variations have produced ecosystems such as peat bogs, sand dunes, grasslands, heaths, moorlands and wetlands. These environments may not only be of local or regional importance, but may also be nationally and internationally significant as conservation sites.

The range of ecosystem types and their classification

There are a number of different approaches to classifying ecosystem types. For example, the Millennium Ecosystem Assessment uses the following groupings:

- **marine** — ocean (where fishing is a major driver of change)
- **coastal** — the interface between sea and land
- **inland water** — permanent bodies of water inland from the coastal zone
- **forest** — lands dominated by at least 40% tree cover
- **mountain** — steep and/or high lands
- **island** — land isolated by surrounding water (high proportion of sea to land)
- **polar** — high-latitude systems frozen for most of the year

- **dry land** — where plant production is limited by water availability
- **cultivated** — land dominated by domesticated plant species
- **urban** — built environments with high human density

A similar classification will be used in this book.

Table 1 lists the ecosystems discussed in the rest of this book. This can be used as a simple framework for study.

Table 1 Topics and case studies covered in this book

Part	Ecosystem/ topic	Spatial extent/ description	Significance	Case Studies
2	Grassland, ecosystems	41% of the land surface (excluding Antarctica and Greenland)	■ Critical producers of protein and fibre from livestock, particularly in developing countries ■ Provide storage of atmospheric carbon ■ Provide recreation and tourism opportunities	1 Tourism in grassland areas, p. 19 2 Mediterranean shrublands, p. 23 3 Desertification in the Sahel, p. 27
3	Forest ecosystems	22% of the land surface (excluding Antarctica and Greenland)	■ Contribute more than 2% of global GDP through the production and manufacture of industrial wood products alone	4 Frontier temperate forests in Chile, p. 36 5 Deforestation in Madagascar, p. 42
4	Coastal ecosystems	Approximately 22% of the land area in a 100-km wide band along the continental and island coastlines	■ Home to approximately 2.2 billion people (about 40% of the world's population) ■ Yields 95% of marine fish catches ■ Home to many threatened ecosystems, e.g. coral reefs and salt marshes, which are ecosystems 'at the edge'	6 Coral reefs as eco-systems, p. 51 7 Seagrasses, p. 56 8 Salt marshes in south-west Wales, p. 58
5	Freshwater ecosystems	Less than 1% of Earth's surface	■ Source of water for drinking, domestic use, agriculture, and industry ■ Freshwater fish and molluscs are a major source of protein for humans	9 Freshwater biodiversity in east Africa, p. 63 10 Mesoamerican wetlands, p. 67
6	Ecosystems of low primary productivity	Include mountain, polar and desert ecosystems	■ Can be the most spectacular, and provide important goods and services to the people who live in such areas	11 Upland ecosystems in England, p. 74 12 Climate change and Arctic wildlife, p. 78 13 The Great Basin Desert, USA, p. 82
7	Human ecosystems	Agro and urban ecosystems	■ Agro-ecosystems provide the over-whelming majority of food and feed on which humanity depends for its continued well-being ■ In 1997, agriculture provided 94% of the protein and 99% of the calories consumed by humans ■ Urban areas occupy only about 2% of the land surface area, but are home to about 3 billion people	14 London's wastelands, p. 86 15 Farming in Cuba, p. 91
8	Biodiversity: loss and conservation	Global — both terrestrial and marine	■ Biodiversity is important for a number of reasons (see p. 94); loss of biodiversity on a world scale is an especially worrying trend ■ Conservationists are attempting to treat the causes of biodiversity loss as well as the degradation of habitats	16 Biodiversity in Shropshire, p. 99 17 Community conservation in Kenya, p. 107 18 Sustainable forest management, p. 109

Understanding how ecosystems work

Ecosystems perform two basic functions: the redistribution of energy and the recycling of nutrients.

Energy flow

Since energy cannot be created or destroyed, it must be transferred from one state to another. The energy from the sun that is fixed by plants during photosynthesis must:

- be stored as chemical energy in plants or animal materials (living or dead), i.e. the biomass; or
- be passed through the ecosystem feeding levels (trophic levels) along food chains and food webs; or
- escape from the system as outputs of material (e.g. decay and excreta or heat energy via respiration).

Figure 8 summarises energy flow through an ecosystem and relates this to biomass produced.

Primary ecological productivity can be defined as the rate at which energy can be converted into organic matter. It can be measured by the amount of new biomass produced each year. Ecological productivity depends on:

- temperature, which speeds up the rate of chemical reactions
- moisture — a key component in many chemical reactions
- nutrient availability from soils etc.
- light, for photosynthesis

Where these factors occur in abundance, for example in equatorial areas or shallow tropical waters, this will lead to maximum primary productivity. High primary productivity supports high **biodiversity**. If any one of the four factors is missing, this will limit overall primary productivity. Clearly, solar radiation is the key controlling factor via insolation. The amount of sunlight reaching an area varies with latitude, altitude, the season and day length.

- **Gross primary productivity** (GPP) is a measure of all the photosynthesis that occurs within an ecosystem.
- **Net primary productivity** (NPP) is the energy fixed in photosynthesis minus the energy lost by respiration (R).

$$NPP = GPP - R$$

NPP is a measure of the new growth available for other levels of the food chain to use. It is measured as a dry weight of biomass. Table 2 shows how NPP and plant biomass vary for different types of land-based global ecosystem (biomes) and global marine ecosystems.

(a)

Energy returned to the system via decomposers and detrivores

Net quaternary production

Examples

Energy used for life processes

Trophic level 4
Tertiary consumers

Energy loss through decay and excreta

Net tertiary production

Energy used for life processes

Trophic level 3
Secondary consumers

Energy loss through decay and excreta

Net secondary production

Energy used for life processes

60–90%

Trophic level 2
Primary consumers

Energy loss through decay and excreta

Net primary production

Energy loss through respiration

40–50%

Trophic level 1
Producers (green plants)

Plant tissue decay; decomposition pathway

Gross primary production

Energy flow

Less than 50% of energy is of wavelengths suitable for photosynthesis

Absorbed by soil surface and vegetation

Total sunlight energy 100%

1.1% of total energy is used in photosynthesis

50% reflected back

(b)

50% energy loss at each level by respiration and decay

Trophic level 4
Trophic level 3
Trophic level 2
Trophic level 1

Global ecosystem	Ecosystem area (million km²)	NPP per unit area (g m⁻² yr⁻¹)		World NPP (billion tonnes yr⁻¹)	Biomass or standing crop (kg m⁻²)		World biomass (billion tonnes)
		Normal range	Mean		Normal range	Mean	
1 Tropical rainforest	17.0	1000–3500	2200	37.4	6–80	45	765
2 Tropical seasonal forest	7.5	1000–2500	1600	12.0	6–60	35	260
3 Temperate evergreen forest	5.0	600–2500	1300	6.5	6–200	35	175
4 Temperate deciduous forest	7.0	600–2500	1200	8.4	6–60	30	210
5 Boreal forest	12.0	400–2000	800	9.6	6–40	20	240
6 Woodland and shrubland	8.5	250–1200	700	6.0	2–20	6	50
7 Tropical grassland (savanna)	15.0	200–2000	900	13.5	0.2–15	4	60
8 Temperate grassland	9.0	200–1500	600	5.4	0.2–5	1.6	14
9 Tundra and alpine	8.0	10–400	140	1.1	0.1–3	0.6	5
10 Desert and semi-desert scrub	18.0	10–250	90	1.6	0.1–4	0.7	13
11 Extreme desert, rock, sand and ice	24.0	0–10	3	0.07	0–0.2	0.02	0.5
12 Cultivated land	14.0	100–3500	650	9.1	0.4–12	1	14
13 Swamp and marsh	2.0	800–3500	2000	4.0	3–50	15	30
14 Lake and stream	2.0	100–1500	250	0.5	0–0.1	0.02	0.05
Total continental	149.0		773	115.2		12.3	1837
15 Open ocean	332.0	2–400	125	41.5	0–0.005	0.003	1.0
16 Up-welling zones	0.4	400–1000	500	0.2	0.005–0.1	0.02	0
17 Continental shelf	26.6	200–600	360	9.6	0.001–0.04	0.01	0.27
18 Algal beds and reefs	0.6	500–4000	2500	1.6	0.04–4	2	1.2
19 Estuaries (includes mangroves)	1.4	200–3500	1500	2.1	0.01	1	1.4
Total marine	361.0		152	55.0		0.01	3.9
Overall total/mean	510.0		333	170.2		12.31	1841

Table 2 NPP and plant biomass for different biomes

Using case studies 1

Question
The graphs in Figure 9 show the relationship between productivity, temperature and precipitation for forest ecosystems. Describe and explain why there is a relationship between (a) temperature and productivity and (b) precipitation and productivity.

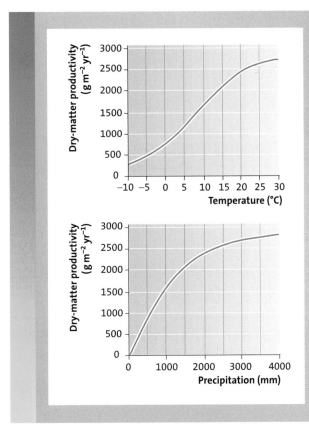

Guidance

(a) Temperature (heat) controls the rate of plant metabolism, which in turn determines the amount of photosynthesis that can take place. Most biological metabolic activity occurs within the range 0–50°C. There is little activity above or below this range. The optimal temperatures for productivity coincide with 15–25°C — optimal range of photosynthesis. The relationship is approximately linear.

(b) Water is a principal requirement for photosynthesis and the main chemical component of most plant cells. In dry regions, there is a linear increase in NPP with increased water availability. In the more humid forest climates of the world, plant productivity begins to level off at higher levels of precipitation.

Nutrient cycling

Nutrients are the chemical elements and compounds required for organisms to grow and function. Nutrient cycling and energy flow are interdependent.

- The rate of nutrient cycling may limit the rate at which energy can be trapped, i.e. plants cannot grow in order to trap more energy if essential nutrients are absent.
- The amount of energy flow may also limit the rate of nutrient cycling that takes place. If a plant is not capturing enough energy for life processes, then its death will result in the breakdown of its organic matter.

Figure 10 shows a simplified nutrient cycle for a tree ecosystem.

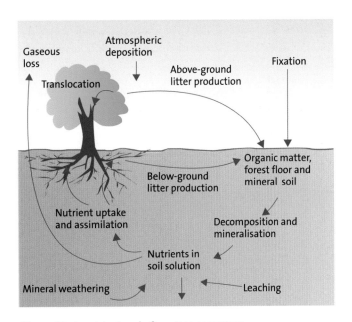

Figure 10 *A nutrient cycle for a tree ecosystem*

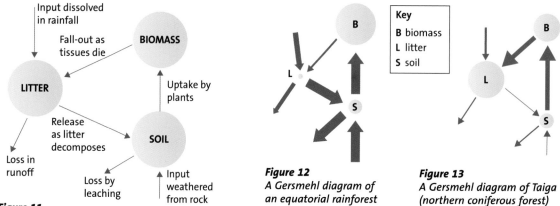

Figure 11
*A generalised
Gersmehl nutrient
cycle model*

Figure 12
*A Gersmehl diagram of
an equatorial rainforest*

Figure 13
*A Gersmehl diagram of Taiga
(northern coniferous forest)*

Figures 11–13 demonstrate how Gersmehl diagrams can be used to illustrate the relative size of nutrient stores in different forest ecosystems. The thickness of the arrows in Figures 12 and 13 indicates the proportion of nutrients transferred. More Gersmehl diagrams can be found in Part 2, Figures 24 and 25 (page 22).

Note the following about an equatorial rainforest (Figure 12):

- the large above-ground biomass, where most nutrients are stored, leads to high biodiversity
- evergreen trees lead to all-year-round leaf fall
- decay is rapid, so there is a low litter nutrient store
- the small nutrient store in soil is a result of very rapid leaching because of high precipitation, which also causes loss by runoff
- there are often deep soils, perhaps 10 m over parent material, but only a few top centimetres are fertile
- there is an active cycle of nutrient circulation, so problems can occur with deforestation as soil is starved of nutrients

Note the following about a Taiga forest:

- the biomass acts as a smaller store of nutrients compared with equatorial rainforest
- cold conditions lead to smaller size of trees, adapted to strongly acidic soils
- the litter layer is the largest store of nutrients, as the humus leads to acidic conditions, which inhibit the breakdown and decay of litter
- the soil is a poor nutrient store — strongly acidic conditions lead to podsolisation, an extreme form of leaching
- overall, this shows a poor nutrient cycle with slow replenishment

Succession

Succession is the gradual and orderly process of change in an ecosystem, brought about by the progressive replacement of one community by another, until a stable climax is established. Figure 14 illustrates an example of succession from open water to oak woodland.

Succession ends with a **climax ecosystem**. For land-based (terrestrial ecosystems), the climax community is often a forest.

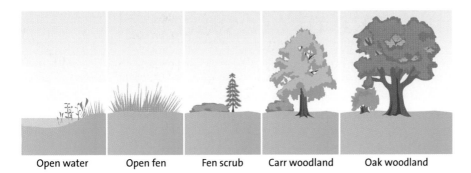

Figure 14
An example of succession

Open water   Open fen   Fen scrub   Carr woodland   Oak woodland

In some cases, repeated disturbances keep an area from ever reaching a climax ecosystem. For example, fires in prairie and chaparral communities keep them from being overgrown by forest. These are often termed **plagio-climax ecosystems**, where succession has been partially deflected, often by human activity. See page 22 in Part 2 (Grasslands) for more on the process of succession.

There are two main types of succession.

- **Primary succession** takes place in an environment that has never been colonised by organisms. An example is the growth of lichens on bare rock surfaces following glaciation.
- **Secondary succession** takes place in an environment that has been colonised before, and where the original inhabitants have never been completely wiped out. An example is the growth of forests on abandoned farmland.

Keystone species are those species whose presence in an ecosystem has major effects on overall diversity. The loss of a keystone species 'ripples' through an ecosystem and causes loss of other species.

The evolution, structure and functioning of ecosystems are increasingly influenced by human activities. Figure 15 shows generalised ecosystem evolution pathways and the effects of human impacts. In reality, there are very few natural ecosystems left anywhere in the world where succession is allowed to continue with no human intervention.

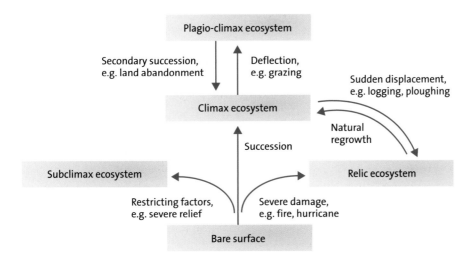

Figure 15
Generalised ecosystem evolution pathways and the effects of human impacts

The state of ecosystems

Perhaps many of the challenges we face today — deforestation, soil erosion, desertification, salinisation and loss of biodiversity — were problems even in ancient times. What is different now is the scale, speed and long-term nature of modern civilisation's challenges to the Earth's ecosystems. The cumulative actions of rapidly growing and industrialising societies have given rise to more complex problems. Acid rain, greenhouse gas emissions, ozone depletion, toxic waste and large-scale industrial accidents are examples of such challenges, with global or regional consequences.

Auditing the current world ecosystem resource

Where are we now? It is difficult to measure the overall condition or health of any particular ecosystem. The ecosystem 'indicators' most readily available, and that have shaped our current understanding of ecosystems, are far from complete. Each provides only a partial description of the bigger picture, rather like the parable of the five blind men giving different descriptions of the same elephant, because each can feel only a small part of the whole animal.

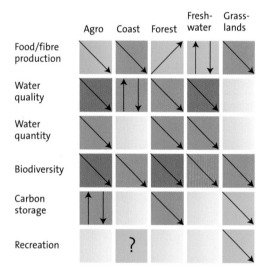

The available indicators include:

- *pressures* on ecosystems, including such factors as population growth, increased resource consumption, pollution and over harvesting
- *extent* of ecosystems — their physical size, shape, location and distribution (using satellite imaging)
- the *production* or output of various economically important goods by the system, such as crops, timber or fish

Each of these indicators is important, but collectively they provide only a narrow view of ecosystem condition and how well ecosystems are being managed. Indicators of pressure, for example, reveal little about the actual health of the system. With proper management, an ecosystem can withstand significant pressures without losing productivity. Indeed, some agro-ecosystems have withstood the pressure of intensive cultivation for generations, but have sustained productivity with the help of organic fertilisers and crop rotation.

The ecosystem scorecard

One approach, commissioned by the World Resources Institute (WRI), is the development of a global 'ecosystem scorecard'. An example is shown as Figure 16.

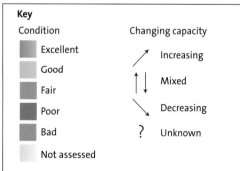

Figure 16 An ecosystem scorecard (for further information, see www.millenniumassessment.org/en/index.aspx)

The **condition** score (indicated by colour) reflects how the ecosystem's ability to yield goods and services has changed over time by comparing the current output and quality of these goods and services with output and quality 20–30 years ago. The condition score is drawn from indicators of production, such as crop harvest data, wood production, water use and tourism, as well as data on biological conditions, such as species declines, biological invasions or the amount of carbon stored in the vegetation and soils of a given area.

The **changing capacity** score reflects the trend in an ecosystem's biological capacity, i.e. its ability to continue to provide a good or service in the future. The changing capacity score integrates information on ecosystem pressures with trends in underlying biological factors such as soil fertility, soil erosion and salinisation, condition of fish stocks and breeding grounds, nutrient loading and eutrophication of water bodies, fragmentation of forests and grasslands, and disruption of local and regional water cycles.

The Living Planet Index

An alternative to the WRI scorecard is 'The Living Planet Index' (LPI — see Figure 17). This attempts to show the overall state of the Earth's natural ecosystems including data on human pressures on natural ecosystems arising from the consumption of natural resources and the effects of pollution. The LPI is formed from an aggregate of three main indicators of the state of natural ecosystems:

- the area of the world's natural forest cover
- populations of freshwater species around the world
- populations of marine species around the world

Figure 17 shows how the LPI fell by 37% between 1970 and 2000. Other LPIs can be found on pages 41, 50 and 66.

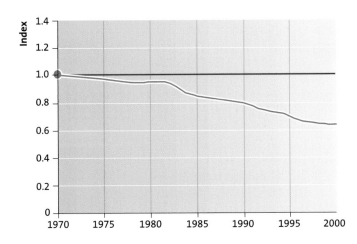

Figure 17
The Living Planet Index, 1970–2000 (1970 is given a value of 1.0)

Using too much?

There is much concern about over consumption and usage of biological resources. About 20% of all biological matter produced by photosynthesis is consumed by humans; as expected, rich people consume more biological resources than poorer

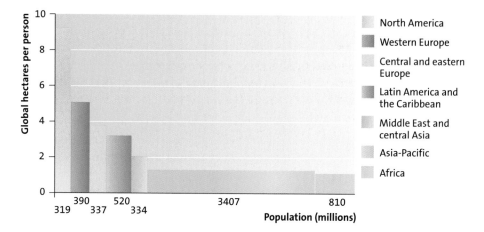

Figure 18 Ecological footprint by region, 2001

people. Most striking is the comparison of consumption patterns with domestic biological productivity. Sparsely populated areas consume less than 1% of the productivity of their lands, whilst large conurbations consume many hundred times more than they produce.

Also of significance is the idea of an 'ecological footprint' – a measure of the mark we leave on the natural world that sustains us (Figure 18). The ecological footprint considers how much land and sea are required to provide us with the water, energy and food we need to support our lifestyles. It helps us judge how sustainable our lives are, and what changes must be made now and in the future to improve our quality of life. The concept was created in the early 1990s and is now in use in many countries at national and local levels.

If the Earth's resources were shared equally among everyone, a 'fair share' would be just less than 2 hectares per person. The UK has a 'footprint' of about 5.5 global hectares per person. This means that if all the world's population had consumption patterns similar to the UK's, we would need two more planets to sustain ourselves. The 'big hitters' of the UK's footprint are materials and waste (38%), food consumption (29%) and direct energy use (18%). A worrying trend is that, on a global scale, our global footprint is increasing continuously and is now unsustainable for the size of the planet, as shown in Figure 19.

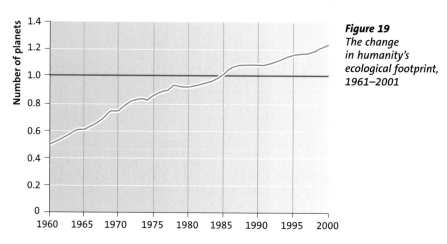

Figure 19
The change in humanity's ecological footprint, 1961–2001

Questions

Look at Figure 20.

(a) What is the difference between a factor and a player?

(b) Why are some players perceived as 'good' and others 'bad'. Is this always the case?

(c) Explain how the conditions of the world's ecosystems are affected by many factors and players.

(d) With reference to one named global ecosystem, assess its value as a provider of goods and services.

International institutions
(e.g. UN agencies, The World Bank, World Trade Organization, etc.)
- direct development aid and funds for investment toward environment-friendly or detrimental activities
- craft and enforce agreements to protect the global environment
- determine trade practices that harm or protect the environment

Figure 20 A world of influences on the environment *(from* World Resources, 2002–2004)

Criminal activities
corruption, illegal logging, and other natural resource theft

Science and technology
pollution control, resource extraction technologies, efficiency improvements

Natural conditions
climate, geography, weather patterns, natural disasters

TNC's/corporations
- determine which goods and services are produced and how (by environment-friendly or detrimental methods)
- drive innovation and technological change
- implicitly or explicitly place values on ecosystem goods and services

Individuals
- demand and use food, fuel, water, shelter
- choose to consume or avoid products that are environment-friendly or damaging
- use ecosystems as places for recreation and spiritual sustenance
- contribute to social norms about environmental behaviour

Political stability
peace, stable financial markets, rule of law, stable bureaucracy

Social and economic conditions
education, poverty, population size, values, religion, culture, distribution of wealth

Governments
- establish and enforce laws that determine who has the right to use the environment and the duty to protect it
- manage natural resources, including collective environmental goods such as clean air and parkland
- decide which environmental uses are taxed and which are subsidised
- restrict corporate and individual behaviours that pose environmental threats
- define and enforce the rules of the market
- designate funds for preservation or development
- redistribute resources between rich and poor

Voice and access
people's ability to organise and take part in decision-making processes

Key

FACTOR

PLAYER

Guidance

(a) Players are individuals or groups of people/organisations arranged at a range of scales, whereas factors are physical, economic and social forces that have an influence on the environment. In reality, it is not always easy to separate factors/forces from players/actors, as there may be strong linkages between them.

(b) Players may be 'good' or 'bad' depending on their impact on the environment. TNCs, for example, are not always bad as they may have sustainable policies in place. However, certain international institutions/governments may suffer from corruption, thereby having an indirect harmful environmental impact. Your answers should be supported with examples.

(c) Use a range of case studies to approach this part of the question. Try to go for a balance of factors and players, together with negative and positive impacts on the environment. It may help to organise your thoughts into a table.

	Positive impact	Negative impact
Player/actor	Use different case studies (at a range of scales) to illustrate the different points.	
Factor/force		

(d) A range of possibilities are available here — more popular choices might include tropical forests and coral reefs. The relevant chapters that follow open with a summary of key goods and services for different ecosystems.

Summary

1 An ecosystem is an open system of organisms functioning together with their non-living environment — the biotic and abiotic components respectively. Some examples are given below:

Biotic components	Abiotic components
Primary producers	Sunlight
Herbivores	Temperature
Carnivores	Precipitation
Omnivores	Water or moisture
Detritivores	Soil or water chemistry (e.g. NH^{4+})

2 Ecosystems receive inputs and lose outputs of nutrients, energy, water and gas.

3 Energy flows through a system. Originating from the sun, energy is trapped by primary producers and transferred to consumers. Energy is lost at each trophic level

4 Organisms form links with each other within the ecosystem — these links can be displayed as food webs. A pyramid of biomass of organisms can be used, with a decrease in the numbers of individuals at each higher trophic level.

5 Ecosystems are self-regulating and controlled by a series of positive and negative feedbacks.

6 Each individual species has its own special adaptations for its role in the community. This role is known as a niche and the adaptations prevent displacement by other species.

Websites

http://www.millenniumassessment.org/en/index.aspx
The Millennium Ecosystem Report. A detailed document detailing the health and loss of bio-diversity from the world's ecosystems.

http://www.wwf.org.uk/researcher/issues/footprint/index.asp
The WWF has some detailed resources about our ecological footprints.

http://www.panda.org/news_facts/publications/key_publications/living_planet_report/index.cfm
Information and data relating to the 'Living Planet' Reports.

http://www.wri.org/
The World Resources Institute (WRI) is an environmental think-tank that researches to find practical ways of protecting the earth and improving people's lives. It is an excellent source of publications and research materials.

Grassland ecosystems

Distribution, goods and services

Grasslands are found on every continent except Antarctica and are often given local names, e.g. steppes (East Europe and Russia), bush (Australia), savanna (Africa), pampas (Argentina), campos (Brazil) and prairies (North America). Figure 6 (p. 4) shows the global extent of these grasslands.

Table 3 *Grassland types of the world*

Grassland type	Area (million km²)	Percent of total land area
Savanna	17.9	13.8
Shrubland	16.5	12.7
Non-woody grassland	10.7	8.3
Tundra	7.4	5.7
World total	52.5	40.5

Pilot Assessment of Global Ecosystems (PAGE)

Figure 21 *Goods and services provided by grasslands*

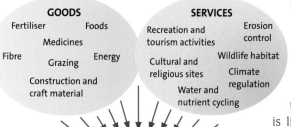

Grasslands are a major biome, covering approximately 40% of the land surface area (Table 3). More people live in grassland areas than in any other ecosystem — approximately 800 million. In Europe alone, around 200 million people inhabit grassland areas. Tundra occurs in the far northern hemisphere, where tree growth is limited by low temperatures and short growing seasons (see Figure 6). It is the least populated grassland type, supporting only 11.5 million people.

Grasslands provide a range of important goods and services (Figure 21).

The importance of grasslands

Grasslands and biodiversity

Grasslands contribute significantly to global biodiversity. They contribute 19% of the world's centres of plant diversity, 15% of the endemic bird areas and 30% of the WWF's unique ecoregions.

Grasslands as drylands

It is important to recognise the idea that many areas of grasslands are drylands, in particular the areas of savanna that are found near the Tropics (see Figure 6, p. 4). Where people are subsistent on these marginal lands (with scant, variable and unreliable rainfall), grassland areas are particularly susceptible to damage by human activities. They are also slow to recover from degradation resulting from mismanagement, such as overgrazing caused by intensification of agriculture to increase food output. The desertification of many semi-arid grasslands is a major global environmental problem.

TOURISM IN GRASSLAND AREAS
Case study **1**

Trophy hunting in Africa

Grasslands are particularly important locations for viewing game animals and for trophy hunting. People are drawn to the large mammals, as well as grassland birds, diverse plant life and generally open-air landscapes. Some recreationalists rely on grasslands for walking and fishing. Others regard specific grassland sites as culturally and spiritually important. Table 4 shows how important tourism is in grassland areas. Figure 22 gives information on the trophy hunting areas in Africa.

Tourism provides important revenues but, at the same time, can be the cause of natural resource degradation. For example, a Wildlife Division Official from Tanzania claimed that:

> Tourist hunters cause less damage than tourists with cameras. The hunters bring less rubbish, cause less damage to roads, use mobile campsites instead of game lodges, and harass fewer wild animals. While trophy hunters may need to locate their target only once, tourists with cameras often are looking for quantity and, in their quest for photos, may drive off roads and follow animals too closely.

The trophy hunting industry:
- attracts attention, both welcome and un-welcome
- provides an emotive activity, for hunters and anti-hunters
- involves a sizeable amount of money
- is a lucrative industry, supporting numerous companies
- requires large capital investments
- provides rural employment
- supports economies in Africa and abroad (air charter, tanneries, curios)
- promotes wildlife use as a viable form of land use
- promotes the country as a tourist destination
- stimulates other tourist-related activities (e.g. photography, hiking)

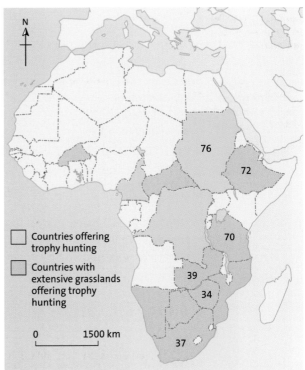

Figure 22 *Trophy hunting areas in Africa, many of which are grassland. The figure within some countries shows the number of different species that can be hunted there.*

Region/country	International tourists (inbound)		International tourism receipts	
	Annual number, 2002 (in 000)	Percentage change since 1985	Annual revenue, 2002 (US$ million)	Percentage change since 1985
Angola	8	–	9	–
Benin	145	150	28	–7
Botswana	693	160	174	361
Burkina Faso	131	181	32	459
Central African Rep	23	331	5	–6
Ethiopia	109	69	30	210
Ghana	305	227	249	801
Guinea	96	–	4	–
Ivory Coast	233	25	78	59
Kenya	703	17	440	46
Madagascar	81	210	64	814
Mozambique	–	–	–	–
Namibia	405	–	214	–
Nigeria	699	202	75	–12
Senegal	287	21	147	33
Somalia	10	–74	–	–
South Africa	4987	621	1962	312
Tanzania	315	199	313	14 441
Zambia	235	85	57	719
Zimbabwe	1722	387	208	624

Source: adapted from WRI

Table 4
The importance of tourism in African countries with extensive grasslands

In Kenya, the quality of national parks and reserves has declined since the 1970s. This is as a result of poorly controlled and excessive tourism and the accompanying increase in the following:

- building of lodges
- water, wood, and electricity consumption
- waste
- off-road driving
- poaching

Although people may be increasing their use of the tourism and recreational services provided by grasslands, the capacity of grassland ecosystems to continue to provide these services appears to be declining.

The ecology of grasslands

The prime factor influencing the distribution of grasslands is climate. Grasslands can be found in areas of low rainfall, or where heat and high rates of evaporation reduce the effectiveness of the rainfall; such areas would be unsuitable for trees. Grasses are well adapted to areas with unpredictable rainfall patterns (see Figure 28, p. 24), as they have a short life cycle and a dense root network. If soils become impoverished in terms of nutrients or rainfall, then areas that are wet enough for forests may become too flooded for grasslands.

Table 5 provides a summary of the similarities and differences between the two major grassland biomes or ecoregions (tropical and temperate).

	Tropical grassland	Temperate grasslands
Approximate latitude	5–20° N and S	40–55° N and S
Climate ■ temperature	22–33°C Variable growing season: 3–8 months Limiting factor: lack of rainfall in winter — dry season Adaptation: rapid life cycle	–15–20°C Average 5 months growing season Limiting factor: low winter temperatures Adaptation: rapid life cycle
■ precipitation	300–1500 mm: concentrated in summer	250–600 mm: late spring/summer maximum Winter precipitation as snow
■ climate type Soils	Tropical: wet/dry Largely ferruginous tropical red earths, includes latosols Incomplete leaching Upward capillary action — concentrates iron	Temperate continental Long-grass prairie: chernozems (black earth) Short-grass prairie: chestnut soils Leaching after spring — snow melt Capillary action — accumulation of calcium
Native species	Numerous game, e.g. antelope, wildebeest	Bison, buffalo
Land use	Traditional herding (Masai) Wild game parks Cash crops: cotton, cattle ranching, peanuts	Traditional herding and hunting, e.g. in Mongolian Steppes (no longer found in North America) Intensive arable farming for cereals/soya beans and ranching
Plant adaptation	Drought-resistant xerophytes in driest savanna (Sahel), e.g. thorn scrub, acacia	Nearly all grass — wind chill from strong winds and physiological drought limits trees

Table 5
The two major grassland types

Grassland productivity

There are also considerable differences in net primary productivity (NPP) between the various grassland biomes. This is largely a function of climate, with the tropical areas experiencing a more suitable climate (warmer and wetter). Refer to Table 2, page 8, for NPP figures.

The two major grassland ecoregions experience some differences in terms of ecosystem structure. This is demonstrated in Figure 23. The key to these ecosystems are the primary producers — green plants capable of manufacturing their own food by photosynthesis (autotrophy). Grass is an example — there are over 9000 species, which are able to grow in a range of environments. Unlike most plants, grass thrives on being eaten as it has multiple growing points or 'meristems'. This is one reason why some grassland ecoregions have productivity rates that are comparable with coniferous tree ecosystems.

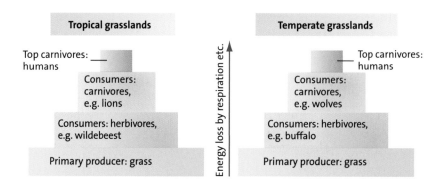

Nutrient recycling

Figure 24
A Gersmehl diagram for tropical grassland

Nutrient recycling is the process by which bacteria and fungi feed off dead and decaying organic matter. During this process, they release certain nutrients essential to plant growth, such as carbon, hydrogen and potassium, which then become available for plant uptake.

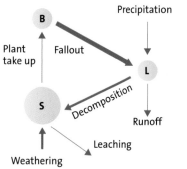

Key
B biomass
L litter
S soil

Figures 24 and 25 illustrate the differences in productivity between tropical and temperate grasslands.

The soil for tropical grasslands contains lower nutrient levels than the soil for temperate ones, with most found in the litter accumulated throughout the year. This is because decomposition is slow during the dry season, when there is very little moisture and the high temperatures inhibit bacterial decomposition. Furthermore, during the wet season, leaching removes nutrients from the soil, which can result in a hard crust forming, hindering plant growth. Biomass varies seasonally, with elephant grasses growing up to 2m high in the wet summer season.

In temperate grasslands, the cold winters and subsequent short growing season mean that the amount of nutrients stored in the biomass is small. Large numbers of bacteria return nutrients from the litter to the soil, which is the largest store of nutrients in temperate grasslands. The relatively dry climate ensures that nutrients are not leached away. The nutrient-rich soil is very fertile, which is the reason why temperate grasslands around the world are farmed.

Figure 25
A Gersmehl diagram for temperate grassland

Issues of succession

Grasslands in both temperate and tropical areas are largely devoid of trees. This is a consequence of certain human factors, such as fire, grazing or use of trees for fuel wood. The result is a plagio-climax of vegetation and not a climatic climax. In grasslands, succession has been 'arrested' (or stopped) by humans and, subsequently, the ecosystem has not been allowed to reach its natural equilibrium (Figure 26). The moorlands of Britain, shown in Figure 27, are an example of a

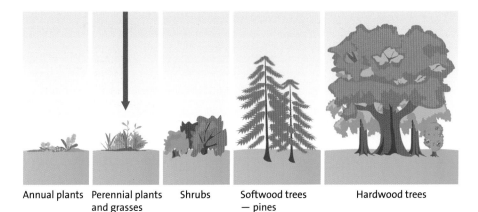

Figure 26
Succession has been arrested at the point shown by the arrow

Annual plants | Perennial plants and grasses | Shrubs | Softwood trees — pines | Hardwood trees

Figure 27
An example of a plagio-climax in the Peak District, UK

plagio-climax. Clearing of the original scrub and tree cover has left an ecosystem that is maintained by burning and grazing — both of which are human influences. The heather habitat shown in Figure 27 is vital to the grouse population. In reality, there is no such thing as a true climatic climax, as humans have already affected every ecosystem on Earth.

MEDITERRANEAN SHRUBLANDS

Case study 2

Mediterranean shrublands are found in five Mediterranean climate regions of the world (Table 6). Their climatic requirements compared with other grassland areas are shown in Figure 28. Mediterranean shrublands are also characterised by regular droughts. Although not always classified as a 'true' grassland ecoregion, they are notable due to their high biodiversity and form a useful case study in terms of human impact on ecosystem dynamics.

Region	Evergreen shrubland	Deciduous shrubland
Southern California	Chaparral	Coastal sage
Central Chile	Matorral	Jaral
South and southwest Australia	Kwongan	Mallee
Southwest South Africa	Fynbos	Renosterveld
Mediterranean basin		
Greece	Xerovini	Phrygana
Spain	Matorral	Tomillar/romero
France	Maquis	Garrigue

***Table 6** Names of Mediterranean shrublands from around the world*

Figure 28 Climatic
requirements
for the various
types of grassland

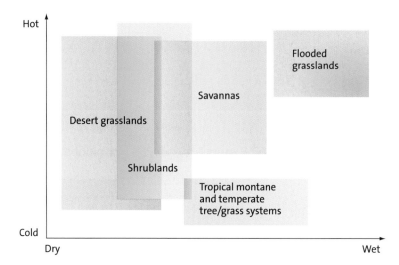

Figure 29 shows how the Mediterranean habitat has been recreated at the Eden Project, Cornwall, UK.

Figure 29
The Mediterranean
'biome' at the Eden
Project

Grazing in the Mediterranean basin

The Mediterranean basin conjures up a number of images: air scented with rosemary; slow pace of country life; shepherds herding their flocks. But these images hide a story of human influence unparalleled in other Mediterranean regions. Forest cutting has long been an impact, but grazing by goats has also been an important influence.

Fire and disaster in southern California

Southern California is prone to fires, the effects of which have become an integral landscape component. Fire is a problem for many homes built in the shrublands. Here, fires have been occurring for millennia and the indigenous shrubs have adapted to cope with this – examples include chamise (*Adenostoma fasciculatum*) California buckwheat

- Seagrass species range in length from the 2–3 cm long leaves of sea vines in deep water off Brazil, to the strands of eelgrass that grow to more than 4 m in the Sea of Japan.
- Seagrass meadows are home to fish, manatees, dugongs and green turtles, and provide a good habitat for many other plants.

Threats

The UN has released the first map of the global distribution of seagrasses. According to the UN, 15% of seagrasses have been lost in the last 10 years. Seagrasses are subject to many threats — human-induced (direct and indirect) and from climate change.

- Direct threats come from boating, land reclamation and other construction in the coastal zone, and from dredge-and-fill activities and destructive fisheries practices (including aquaculture).
- Indirect threats come from runoff of nutrients and sediments from human activities. Nutrient and sediment loading affect water clarity; seagrasses have relatively high light requirements, which makes them vulnerable to decreases in light penetration of coastal waters.
- Global climate change may well impact on seagrass distribution, as sea levels rise and storms occur more frequently.

Widespread areas of seagrass loss have proved difficult to restore across the world. There have been few successful rehabilitation exercises, often because the original cause of loss has not been ameliorated. Where environmental damage has been serious, little seagrass recovery has occurred. Re-vegetation efforts are expensive, but have proved effective on small scales, particularly in sheltered environments.

Improving the seagrass habitat — Gulf of Mexico

Seagrass degradation and loss of habitat in the northern Gulf region (including the states of Texas, Florida, Mississippi, Alabama and Louisiana) have had significant impacts on natural resources and socioeconomic vitality. However, with the right approach, it is possible to reverse the current trends of seagrass loss.

- Management plans are focusing on research, monitoring, conservation, restoration, protection and education, facilitating discussion and the prioritisation of major issues.
- Promotional literature and informational websites are used to communicate with the general public.
- Monitoring is the cornerstone of the programme. The *Status and Trends Report* represents an initial effort to create a baseline of information and increase stake-holder awareness.

9 Question

Using case studies

With reference to marine conservation areas, discuss some of the problems experienced with their management.

Guidance

Here it is best to go for a range of marine ecosystems, as well as different locations. Perhaps examples can be chosen where there are different types of management, for example protectionist strategies and lower-impact conservation measures. Problems are likely to be generic in many instances, for example pollution, climate change, over-fishing, impacts of tourism etc. However, problems may also exist locally and be unique to that particular place.

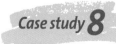

Complex and fragile ecosystems

Salt marshes are some of the most threatened ecosystems on the planet, as they occupy coastal locations that are attractive for industrial, commercial and recreational use. These uses create pressures for development and contribute to pollution and other damage to the marshes. The marshes are also vulnerable to environmental change. This includes global climate change, which is manifesting in rising sea levels.

In the past, salt marshes have not always attracted much sympathy as candidates for protection. They lack the spectacular scenic attractions of cliffs and fjords or the sunbathing opportunities of sandy beaches.

Formation of salt marshes and succession

Salt marshes occupy a mid-level between the mudflats, which edge the permanently submerged marine zone, and the land-based (terrestrial) habitats, which lie above the high water mark (Figure 62).

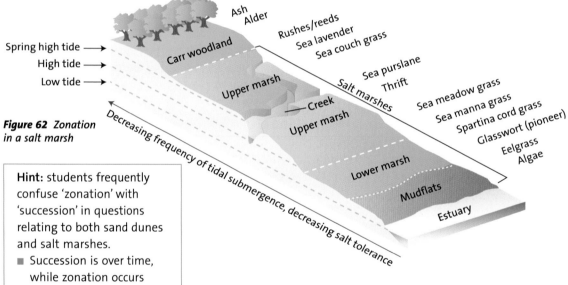

Figure 62 Zonation in a salt marsh

Hint: students frequently confuse 'zonation' with 'succession' in questions relating to both sand dunes and salt marshes.

- Succession is over time, while zonation occurs over space.
- Zonation can be easily measured in the field. Succession happens over a long time period and it is hard to provide evidence for it.
- Figure 14 (in Part 1) shows an example of succession.
- Figure 62 shows zonation.

Like sand dunes, salt marshes result from the interaction of geomorphological and ecological processes; however, the development of salt marshes is restricted to low-energy environments, where the sea's erosional energy is limited.

The Gann Flats, Pembrokeshire

The Gann Flats at Dale, Pembrokeshire (west Wales) form a muddy beach with a variety of flora and fauna. This salt marsh SSSI has developed in very sheltered conditions at the mouth of the Afon Gann, behind a shingle storm bar. Its muddy flats, occupying an area of 60 000 m² at extreme low water, slope gently to a height of around 2 m above the river and are characterised by scattered saltpans and creeks.

Site number	1	2	3	4	5	6	7	8	9
Height (cm)	0	25	50	75	100	125	150	175	200
Percentage frequency (not all species listed)									
Bare ground	70	68	14	8	0	0	0	0	2
Green algae	30	31	4	6	0	0	0	0	0
Puccinellia maritima	0	26	58	69	76	38	2	0	0
Halimione portulacoides	0	0	12	9	42	33	28	21	0
Limonium humile	0	0	0	0	0	7	0	4	0
Festuca rubra	0	0	0	0	0	13	18	2	96
Total number of different plant species	1	4	6	8	8	11	10	9	2
Simpson's Diversity Index	1.00	2.65	2.70	2.89	3.98	7.83	6.73	6.12	1.1
pH	8.62	8.57	8.03	8.01	8.39	7.79	7.68	7.44	6.98
% organic matter	7.07	8.50	11.08	12.14	20.00	20.61	21.88	33.36	41.61
Salinity (mV)	−0.058	−0.063	−0.06	−0.051	−0.086	−0.084	−0.05	−0.041	−0.019

Table 18 shows fieldwork data for the Gann Flats. Note that site nine is the highest survey point. Results shown are percentage frequency of species.

Table 18 Field data from Gann Flat

Threats to salt marshes

Salt marsh environments are particularly susceptible to damage due to the delicate nature of their abiotic and biotic relationships. Figure 63 shows the relationship between abiotic and biotic components within a salt marsh ecosystem.

Threats to salt marsh ecosystems include:
- industrial pollution, which can harm marsh species
- agricultural pollution, leading to eutrophication
- shipping and recreational activities which cause 'wash', leading to die-back of vegetation
- pressure from development, i.e. marina and recreational facilities

Other threats (which may be linked to climate change) include:
- increased incidence of high-impact storms, causing habitat destruction
- changing temperature and rainfall patterns, which may affect the tolerance of marsh species
- sea level rises, which may occur too quickly for the marsh environment to adjust

In summary, salt marshes are intertidal habitats comprising salt-tolerant vegetation. The frequency and duration of tidal inundation determines which plants and animal species are present. Salt marshes are bisected by meandering creek systems, which allow tidal waters to drain in and out. The creeks slow down tidal energy and the marsh plants slow down wave energy. There are a range of threats to this delicate ecosystem, including pollution and pressure from development at the coastal fringe.

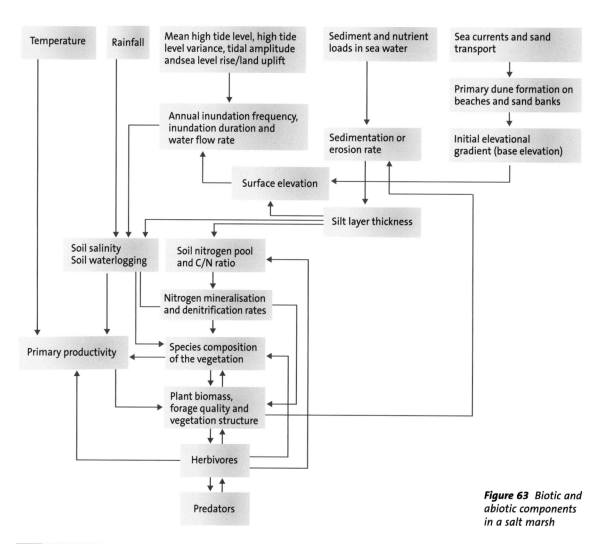

Figure 63 Biotic and abiotic components in a salt marsh

Question

(a) Use a diagram or timeline to show how the vegetation changes during the succession process in a salt marsh.

(b) Design a fieldwork investigation to collect information about the zonation of plant species on a salt marsh. Describe what techniques you might use to present and analyse the data.

(c) Use internet, book and magazine evidence to assess the direct and indirect threats to the survival of salt marshes.

Guidance

(a) Most textbooks have diagrams that show the process of succession and which can be adapted to fit a salt marsh. Examiners are keen for you to demonstrate a clear understanding of the difference between zonation and succession. Note that a discussion of succession must refer to changes within the community over time — it is directional and leads to a stable point, which is the climax community. A discussion of zonation must refer to changes in space (see Figure 62).

(b) You will most likely use a series of transects, coupled with systematic/stratified sampling. Equipment may comprise tapes and quadrats (either frame or 'point'), along with identification charts. Data from this type of survey are often presented in the form of kite diagrams (see Figure 64). Spearman's Rank might be used to analyse the data statistically, in which case 'null' and 'alternative' hypotheses will have to be developed.

(c) 'Assess' means 'weigh up'. Be careful to distinguish between direct threats (e.g. dredging, pollution, building, reclamation), indirect threats (e.g. climate change) and knock-on changes (e.g. from the construction of coastal defences leading to erosion of a spit protecting a salt marsh). An assessment might also look at the role of protection strategies that mitigate the threats.

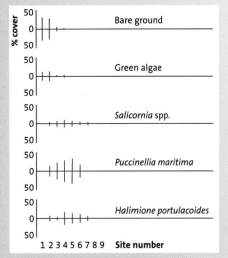

Figure 64 *A skeletal kite diagram of an interrupted transect on a salt marsh*

Summary

1 Coastal ecosystems, found along continental margins, are regions of remarkable biological productivity and high accessibility.

2 Coastal ecosystems provide a wide array of goods and services: they host the world's primary ports of commerce; they are the primary producers of fish, shellfish and seaweed for both human and animal consumption; and they are also a considerable source of fertiliser, pharmaceuticals, cosmetics, household products and construction materials.

3 Coastal areas encompass a broad range of habitat types and provide a home to a wealth of species and genetic diversity. Coastal ecosystems store and cycle nutrients, filter pollutants from inland freshwater systems and help to protect shorelines from erosion and storms.

4 The beauty of coastal ecosystems makes them a magnet for the world's population. People gravitate to coastal regions to live, as well as for leisure, recreational activities and tourism.

5 Human development is threatening many coastal ecosystems, but human activity is also restoring some previously damaged coastal environments.

Websites

http://marine.wri.org/pagecoastal-pub-3054.html
WRI assessment of coastal ecosystems.

http://www.millenniumassessment.org/en/Article.aspx?id=76
The Millennium Ecosystem Assessment on coasts and marine areas.

http://www.seagrasses.org/
Detailed information about seagrasses and their habitats.

http://www.unesco.org/csi/pub/papers/papers3.htm
The Caribbean Coastal Marine Productivity program (CARICOMP).

Freshwater ecosystems

Distribution, goods and services

Freshwater systems are created by water that enters the terrestrial environment as precipitation and flows both above and below ground towards the sea. These systems encompass a wide range of habitats, including rivers, lakes and wetlands, and the waterside zones associated with them (Figure 65). Their boundaries are constantly changing with the seasonality in the hydrological cycle.

The goods and services derived from inland waters have an estimated global value of several trillion US$. Goods include:

- **water** (freshwater systems provide water for human, agricultural and industrial use, as well as for recharging aquifers)
- **fish** (in 1997 the catch from rivers, lakes and wetlands was 7.7 million tonnes and the yield from aquaculture was 17.7 million tonnes)
- **energy** (of the total amount of electricity generated globally, hydropower from water stored in dams accounts for 18%)

Services include:

- **biodiversity** (of all animal species, 12% live in freshwater; many others, including humans, depend on water for survival)

Figure 65 *River systems provide a range of goods and services*

David Holmes

- **waste removal** (rivers and streams, together with the plant life around them, filter pollution and maintain water quality)
- **recreation** (fishing, boating, swimming and bird watching at freshwater sites provide employment and tourism opportunities)
- **flood control** (streams, rivers and their catchments can help regulate and buffer against flood risks)

The importance of freshwater ecosystems

Species richness, in relation to area of habitat, is extremely high in many freshwater environments. Freshwater fishes and molluscs comprise 40% of all fishes and about 25% of all molluscs, respectively. An estimated 11 000 fish, 5000 amphibian and 6000 mollusc species depend on freshwater habitats. Other major groups dependent upon freshwaters include reptiles, insects, plants and mammals. As freshwater ecosystems occupy only 0.8% of the Earth's surface, this translates into a considerably species-rich group of habitats.

There are some worrying facts, however.

- More than 20% of the world's known 10 000 freshwater fish species have become extinct or threatened in recent decades.
- 'At risk' rates are disproportionately higher for freshwater plants and animals than for terrestrial plants and animals.
- Half of the world's wetlands are estimated to have been lost during the twentieth century.
- Historical trends in data regarding commercial fishing from the major rivers of the world show dramatic declines during the twentieth century.

FRESHWATER BIODIVERSITY IN EAST AFRICA

Case study 9

An example of catchment protection

The inland waters of east Africa are internationally recognised for their high levels of species richness and endemism, particularly within the Rift Valley lakes. The freshwater fisheries supported by this diversity are almost entirely small-scale, providing income and food security to a large portion of the poorer communities. In Malawi, for instance, it is frequently stated that 70% of dietary animal protein is derived from fish, and the fishery sector is thought to constitute a major source of income and livelihood for more than 300 000 people. Over 1 million people are dependent upon the fisheries from Lake Tanganyika.

Disruption of the fisheries in Lakes Tanganyika and Victoria, through loss of biodiversity, has been attributed to over-fishing, eutrophication and the introduction of alien species. This has already led to significant loss of jobs, livelihoods and food security. Fragmentation of habitats and the diversion of water resources for development of hydroelectric power present additional threats to freshwater biodiversity. East Africa is already heavily reliant upon hydroelectricity — 57% of the electricity for the region in 2001 was provided by hydroelectric power.

Figure 66
East Africa and the Rift Valley lakes

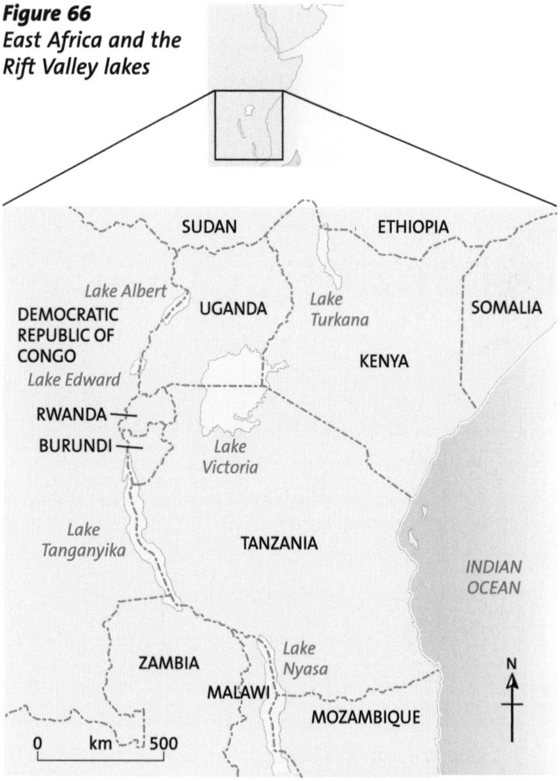

SUDAN
ETHIOPIA
Lake Albert
DEMOCRATIC REPUBLIC OF CONGO
UGANDA
Lake Turkana
SOMALIA
Lake Edward
KENYA
RWANDA
BURUNDI
Lake Victoria
Lake Tanganyika
TANZANIA
INDIAN OCEAN
ZAMBIA
Lake Nyasa
MALAWI
N
MOZAMBIQUE
0   km   500

Conservation of freshwater ecosystems

In general, aquatic conservation strategies should support sustainable development by protecting biological resources in ways that will preserve habitats and ecosystems. In order for biodiversity conservation to be effective, management measures must be broad-based. This can be achieved through many mechanisms (see Figure 67).

The most effective management unit for conservation of freshwater organisms is the river catchment. Inclusion of a species distribution within a protected area does not, in itself, ensure effective protection of the associated wetland habitat, especially if the quality or quantity of incoming water flow is not also maintained through protection of river basin headwaters.

A number of sites have been identified as regionally important for species richness and endemism, and for threatened species. The main sites are the African Great Lakes – Malawi (or Lake Nyasa), Tanganyika and Victoria. The Rufiji, Pangani and Tana River systems have also been identified as important centres for freshwater biodiversity.

Figure 67
Aquatic conservation strategies

Restoration/mitigation efforts
Aquatic areas that have been damaged or suffered habitat loss or degradation can be restored

Aquatic Diversity Management Areas (ADMAs)
The primary goal is to protect the aquatic biodiversity in a given area

Increase public awareness
This is one of the most important ways to conserve aquatic biodiversity

Legislation
(e.g. international or national)

Research
Various organisations and conferences help to identify areas of future research, analyse current trends in aquatic biodiversity, and even conduct specialised studies

Local watershed groups
Rivers and streams often go unprotected if they pass through more than one political jurisdiction, making it difficult to enforce conservation and management of resources; local watershed groups can help in this situation

AQUATIC CONSERVATION STRATEGIES

Threatened or endangered species designations
Once species are 'listed', they become subject to national recovery programmes and are placed under international protection

Bioregional management
Bioregional management is a total ecosystem strategy, which regulates factors affecting aquatic biodiversity by balancing conservation, economic and social needs within an area

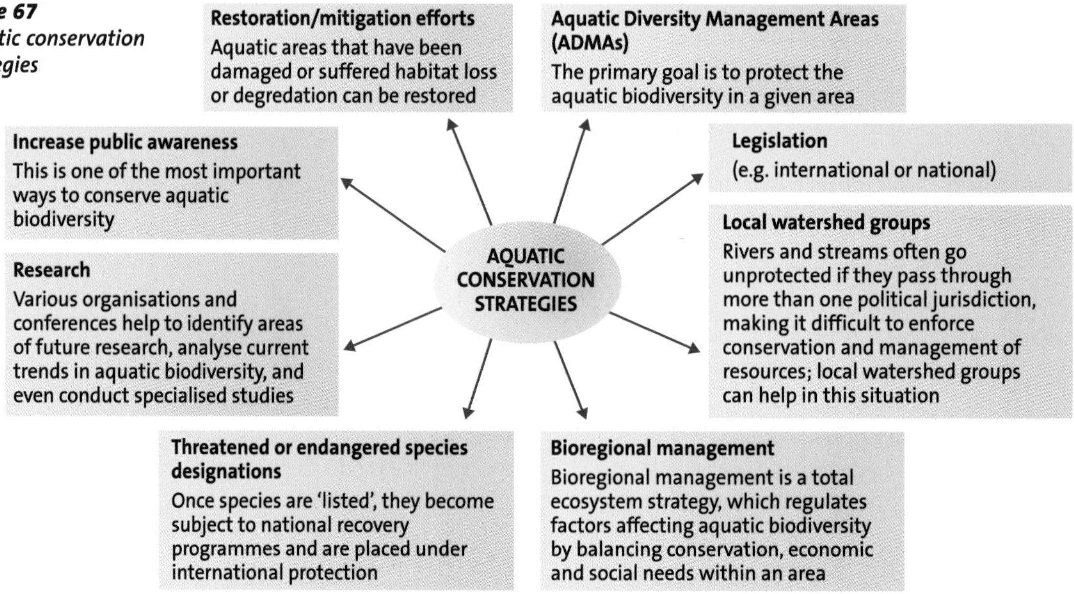

An additional major threat recognised as common to all the taxonomic groups is the continuing loss of habitat through deforestation and agricultural encroachment. Forest reserves, which often aim to protect forests in the upper river catchments, are highlighted as potentially important tools in need of increased recognition for their value to the protection of downstream freshwater ecosystems.

Work is ongoing to finalise the site selection criteria used to identify Key Biodiversity Areas (KBAs) for inland waters. This work is part of a broader collaborative initiative, to identify KBAs for terrestrial and marine ecosystems at the global scale, led by Conservation International and BirdLife International.

Using case studies

11 Question

'Throughout the UK, raised bogs have suffered dramatic losses over the last few centuries, largely due to drainage and conversion to agriculture and forestry.' (*Out of the Mire*, RSPB) Examine the ways in which the rate of loss of raised bog can be slowed or even halted.

Figure 68
Peat extraction in western Ireland

David Holmes

Guidance

A range of strategies can be used. The first few are to do with site conservation:

- International legislation: the issue of raised bogs is included in the Habitats and Species Directive (92/43/EEC). The ecosystem is given special status as one of the most threatened habitats in Europe.
- UK wildlife legislation: statutory nature conservation agencies (e.g. Natural England) can notify peat bogs as SSSIs, which offers some levels of protection.
- Management agreements: statutory agencies have powers to enter management agreements with developers.
- Rehabilitation: this may include planting particular species, clearing woods etc. It is about positive management of bog landscapes.

- Land-use planning: local authorities can review and revoke existing planning permissions for peat extraction, impose strict development controls and also restrict expansion of agriculture and forestry.

Other strategies are used to replace peat in horticulture:

- Peat-free products: there are a range of peat-free alternatives on the market; EU Eco-Labelling regulation ensures better consumer information and encourages retailers to stock less environmentally damaging products.
- Use of organic wastes: the switch from peat gives an opportunity to recycle more organic products, e.g. at home and on a larger scale. This includes coconut fibre pitch (coir), bark and animal manure.

Threats to freshwater ecosystems

Despite their clear economic value, many inland water ecosystems, especially wetlands, have long been considered as a wasteful use of land and are rarely protected. Lack of recognition for the value of these systems has already led to the loss of many of the world's wetlands. Rates of species loss have, in some cases, been estimated at five times greater than those seen in other ecosystems. The Freshwater Species Index (Figure 69) shows a decline of approximately 50%, from 1970 to 2000, in 323 vertebrate species found in rivers, lakes and wetland ecosystems.

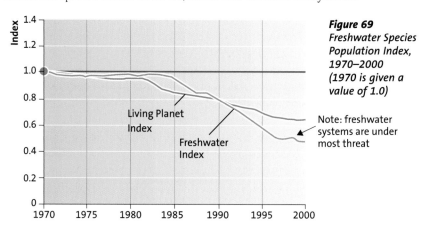

Figure 69
Freshwater Species Population Index, 1970–2000 (1970 is given a value of 1.0)

The main causes of freshwater biodiversity decline are:
- hydrologic alteration from dams and diversions
- water quality degradation, largely from agriculture, including eutrophication
- predation and competition from invasive species
- climate change

Over the last 100 years, the scope of these threats has increased hugely, alongside a six-fold increase in our appetite for the range of services provided by freshwater globally (transportation, irrigation, energy production etc.)

Other areas for concern include the following:
- About 60% of the world's larger rivers are now moderately-to-severely altered by dams and diversions. Today, there are more than 45 000 'great' dams in the world, most of them in China and the US, but increasing numbers of new dams are now being built in Asia and South America.
- Worldwide, water quality conditions are deteriorating in almost all regions recently assessed by the World Resources Institute. Intensifying agriculture and urbanisation, along with inadequate sanitation systems, are significantly affecting the world's rivers, lakes and wetlands.
- Non-indigenous invasive fish and plants threaten native species through predation, over competition for limited resources and habitat degradation. Invasive species can out-compete native species, allowing them to replace native stock in rivers, lakes and wetlands.
- Over-extraction of water from precious groundwater supplies is leaving many rivers and lakes in a low-quality state. In addition to increasing demand from

domestic supplies, agriculture, bottled water purveyors and industry are also increasing their rates of abstraction.

MESOAMERICAN WETLANDS

Central America and Mexico

Mesoamercia is a huge land area which encompasses Mexico in the north and Panama in the south. More than half of Mesoamerica's wetland sites are marine-coastal and include mangroves, estuaries and coral reefs. Rivers, however, are the arteries of Mesoamerica. They are of vital importance, providing water for human consumption, industry and agriculture. They also offer a means of transportation, recreation and tourism, as well as a rich source of biodiversity.

Unfortunately, the rivers and lakes of Mesoamerica are increasingly threatened by farming and industrial activities, as waste and sediments are emptied into these valuable ecosystems. The rivers and most of the largest watersheds are being damaged by deforestation, erosion and pollution.

The Colorado River is an example of a Mesoamerican river that has been heavily altered by humans. The Hoover Dam (plus a myriad of other dams), diversion canals and channels, have converted much of the river to little more than a conveyance canal for the delivery of water to sprawling southwestern cities and big agribusiness. Over years, the US government has signed contracts and a treaty for the delivery of water to river basin states (Arizona, California, Colorado, Nevada, New Mexico, Utah and Wyoming) and to Mexico, for far more water than actually flows downstream in any given year! Figure 70 shows the impact on discharge levels.

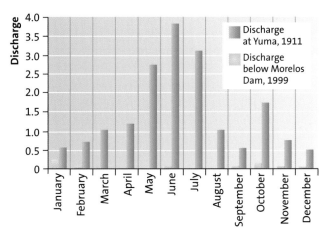

Figure 70 *Seasonal variation of river flow at Yuma, Arizona, in 1911 (before the dams) and below Morelos Dam, Mexico, in 1999, below the last diversion*

This complex of natural and artificial wetlands is one of the largest hydrographic basins in North America. Native endangered animals include *Totoaba macdonaldi* (a fish), *Cyprinodon macularius* (a freshwater fish native to the region) and *Phocoena sinus*, a porpoise species endemic to Mexico and one of the rarest in the world.

The mouth of the Río Colorado was designated a reserve by the Mexican government in the 1930s. The reserve covered only the saltwater portion of the delta, for the protection of important fish species. In June 1993, the federal government established the Reserva de la Biosfera Alto Golfo de California y Delta del Río Colorado, with a core area of 164 780 hectares. This reserve is in the north of Mexico (Figure 71).

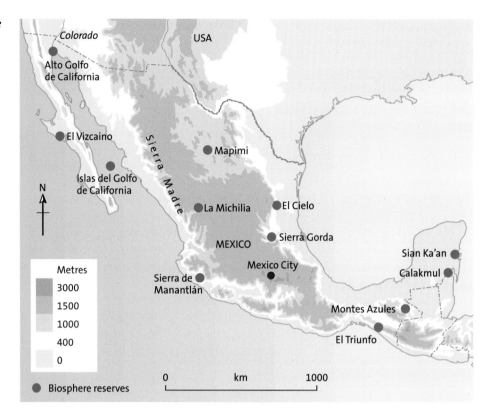

Figure 71 Biosphere reserves in Mexico

The core area of the Río Colorado delta covers the area at the mouth, part of the main channel, the eastern floodplain and the Santa Clara and El Doctor swamps. A management plan has been approved and is being implemented. In 1994, the delta of the Río Colorado was included in the Western Hemisphere Shorebird Reserve Network. The site is also designated as a UNESCO Biosphere Reserve. Proposed conservation measures for the site include construction and operation of the Golf de Santa Clara field station, signage, security and implementation of management plans for proposed hunting reserves. Environmental education programmes and the promotion of ecotourism have also been proposed.

The range of freshwater ecosystems

Lakes, ponds and reservoirs

Lakes are inland bodies of freshwater, ranging in size from less than one hectare to thousands of hectares. They are the bodies of water that fill depressions in the Earth's surface. Lakes may be further described by their origin and classified by trophic status and characteristics (Table 19). Reservoirs are lakes, often man-made, that control water flow for hydroelectric power generation, flood control and/or municipal water supplies. Table 20 compares lakes and ponds.

Lake trophic classification	Nutrient concentration	Biological productivity
Oligotrophic	Low	Low
Mesotrophic	Moderate	Moderate
Eutrophic	High	High
Hypereutrophic	Very high	Very high

Table 19 *Lake trophic classification and characteristics*

Lakes	Ponds
Larger	Smaller
Deeper	Shallower
May have dramatically different temperatures at the surface compared with the bottom	Top and bottom waters generally have the same temperature
Affect local climate if large enough	Tend to be greatly affected by local climate

Table 20
Comparison of lakes and ponds

The importance of, and threats to, lakes

Lakes hold almost 90 % of the liquid surface freshwater on earth and are major regulators of global carbon, nitrogen and phosphorus cycles. They are important reservoirs for freshwater, purifiers of terrestrial waste, and zones for aquifer recharge.

Lakes provide an important habitat for fish, crustaceans, molluscs, reptiles, amphibians, birds, mammals, insects, and aquatic plants. They also support biodiversity on the surrounding land.

The maintenance and biodiversity of lakes is central to the lives of the human populations surrounding them. Lake Manzala, the largest of the Nile delta lakes, produces 30% of Egypt's fish yield, and supports 120 000 fishermen. Lake Nyasa fisheries supply 50% of the animal protein in Malawi.

In many countries, lakes supply a large proportion of the drinking water; Lake Chapala, in Mexico, is the main water source for Guadalajara, where the population exceeds 4.5 million. The lake has been shrinking dramatically due to water diversion for irrigation.

Lakes worldwide are being threatened. The diversion of water for irrigation and for use in industry, invasion by exotic plant and animal species, contamination by toxic substances and nutrients from industry, farms and sewage, together with urban runoff, are all common on a scale today that significantly threatens lake ecosystems on every continent except Antarctica.

Fragmented and poor lake resource management is affecting waters in both developed and developing countries. In most parts of the world, human impacts on lakes are spreading and becoming more intense due to human population increases and the globalisation of trade, which has increased deforestation and the use of pesticides and fertilisers, and has encouraged the spread of invasive species.

Rivers and streams

Rivers and streams are the general terms used to describe both natural and artificial bodies of moving water. Rivers are bigger than streams and empty into large water bodies such as oceans and lakes.

There are three main types of stream:
- Ephemeral streams exist regularly for short periods of time, usually during a rainy period, and may have defined channels even when they are dry.

Figure 72
The River Severn, near Ironbridge, Shropshire

- Intermittent streams flow at different times of the year, or seasonally, when there is enough water from either rainfall, springs, or other surface sources such as melting snow or even discharge from a wastewater treatment facility.
- Perennial streams are those that flow all year round.

Wetlands

A wetland is an area where water is the primary factor controlling the immediate environment. Wetlands can be as small as a child's paddling pool or as large as a lake.

Wetlands generally occur where land and water meet and underground water is at or near the surface, or where land is covered in water less than 1.5 m deep.

Table 21 gives examples of different types of wetland.

Table 21 *Different types of wetland*

Type	Description
Freshwater marshes	- Characterised by periodic or permanent shallow water, little or no peat deposition and mineral soils - Typically derive most of their water from surface waters, including floodwater and runoff, but do receive groundwater inputs too
Wet meadows	- Commonly occur in poorly drained areas such as shallow lake basins, low-lying depressions and the land between shallow marshes and upland areas - Precipitation serves as their primary water supply, so they are often dry in summer
Vernal pools	- Have either bedrock or a hard clay layer in the soil that helps keep water in the pool - Covered by shallow water for variable periods from winter to spring, but may be completely dry for most of the summer and autumn

Wetlands are very important, not only in terms of their biodiversity, but also as flood control measures (acting as giant sponges), silt-catchers and water cleaners.

The edge of any productive water body will show different zones of vegetation, according to the depth of water. A gently shelving edge will have wider and more complex zones than one that drops steeply to deep water.

Examine a productive pond over the years and it becomes clear that each of the vegetation zones shown in Figure 73 gradually moves towards the middle through the process of succession. Rooted plants trap ooze between their stems; when they die and decompose, they further thicken the bottom sediments, which build up and reduce water depth. This in turn allows other plant species to invade and cause further drying; what is at present a few feet of open water covered in water lilies may in a matter of decades become a reedswamp, dry for part of the year; what was once a reedswamp may now be covered in willow scrub or alder carr; what is now deep scrubland may in time become a young pine or oak wood.

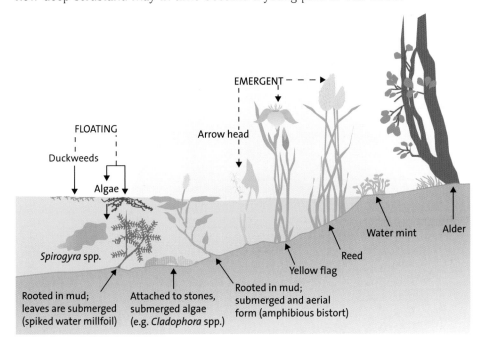

Figure 73
Vegetation zones in a productive pond

EMERGENT

FLOATING

Arrow head

Duckweeds

Algae

Water mint   Alder

Spirogyra spp.

Reed

Yellow flag

Rooted in mud; leaves are submerged (spiked water millfoil)

Attached to stones, submerged algae (e.g. *Cladophora* spp.)

Rooted in mud; submerged and aerial form (amphibious bistort)

Summary

1 Freshwater ecosystems in rivers, lakes and wetlands contain just a fraction (one-hundredth of 1%) of the Earth's water and occupy less than 1% of the Earth's surface.
2 There is an extraordinary range of different types of freshwater ecosystem, with a disproportionately large amount of biodiversity.
3 The most important goods and services that humans derive from freshwater systems revolve around water supply: providing a sufficient quantity of water for domestic, agricultural and industrial use; maintaining high water quality; and recharging aquifers that feed groundwater supplies.

4 With population growth, industrialisation and the expansion of irrigated agriculture, demand for all water-related goods and services has increased dramatically, straining the capacity of freshwater systems.

5 Human activities have severely affected the condition of freshwater systems worldwide. Even though humans have increased the amount of water available for use with the construction of dams and reservoirs, more than 40% of the world's population live in conditions of water stress.

Websites

http://water.wri.org/
The WRI gives details on freshwater ecosystems.

http://legacy.ncsu.edu/classes/nr400001/gradpage/Wetland_Mitigation_Home/
wetland_importance.html |
This gives some details on the importance of wetland habitats.

http://www.worldlakes.org/
An organisation working on the conservation and sustainable management of lake resources.

http://www.epa.gov/bioindicators/aquatic/freshwater.html
An American site, which details some threats to freshwater ecosystems.

Ecosystems of low primary productivity

Low productivity ecosystems are places round the world where, for many organisms, survival is at the edge. Such areas typically have harsh and stressful environments, caused by climatic extremes. They usually have low levels of diversity, but are still important areas for discussion. They include mountain, polar and desert ecosystems.

Mountain ecosystems

Mountain ecosystems are perhaps the grandest and most spectacular of all. Weathering processes and gravity constantly pull rocks, soil, snow and water downhill, inhibiting the development of soils. Thin soils and unstable slopes, in turn, limit plant growth, raise the vulnerability of mountains to human disturbance, and require lengthy recovery time once damaged. Mountain regions also have a long history of political neglect and economic exploitation.

Millions of people who live far from any mountains benefit from a range of goods and services, including water, timber, rich biodiversity and awe-inspiring scenery, that mountain ecosystems supply. Those who live in the mountain and upland regions themselves, about 10% of the world's population, often depend on these mountain ecosystems for subsistence. Within mountainous regions of developing countries, transport links may be scarce, access to supplies and markets poor, population growth rates high, and employment opportunities limited. Mountain populations in Nepal, Ethiopia and Peru, for example, rank among the world's poorest.

The definition of a mountain region can be based on several criteria, including height, slope, climate and vegetation. A simple definition is 'areas above 3000 m', a category that encompasses about 5% of the world's terrestrial surface and an estimated 120 million people. Twenty seven per cent of the world's surface is defined as 'upland' — areas above 1000 m. About 500 million people worldwide live in uplands and mountains.

Importance of mountain ecosystems

Mountain ecosystems cover a range of shapes, climates and compositions of plant and animal species. These all depend on elevation and latitude.

Tourism threats

For a long time, tourism was seen as a benign economic stimulus for mountain communities, unlike mining and grazing, which have demonstrable impacts on the environment. But we are now seeing that tourism comes with its own set of environmental problems.

Records show that since 1985, the number of visitors to the Rocky Mountain National Park in Colorado has grown by 50%. In another example, 20 000 cars a day drive through Canada's national park in Banff, Alberta, at peak season.

In Europe, two-season tourism has grown exponentially in the Alps since the 1950s. Despite warmer-than-usual weather and shorter ski seasons at lower elevations, millions of tourists and their vehicles, are adding to already heavy commercial traffic, causing dangerous air pollution levels in many alpine valleys. In addition, the threat of water pollution stemming from development of all kinds, including mass tourism, is growing in the Alps.

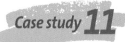

Case study 11   UPLAND ECOSYSTEMS IN ENGLAND

Between a rockery and a hard place

The uplands of England are considered to be some of our last remaining wild areas. Their rugged inspirational landscapes give us a sense of naturalness and space. Yet these areas are far from being true wilderness. What we see today is a landscape that, although the product of powerful geological and biological processes, has also been greatly shaped by centuries of human activity.

Figure 74
Limestone pavement locations in northern England

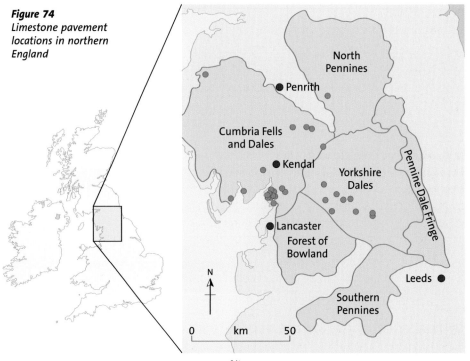

● Areas of limestone pavement

'Limestone pavements' are large areas of rock, scoured by glaciers during the ice ages and then weathered over thousands of years. They contain complex patterns of deep crevices known as grikes, between which are massive blocks of worn limestone, called clints. Limestone pavements are scarce and non-renewable, and Britain and Ireland hold the most important and extensive areas in the world. Figure 74 shows the locations of limestone pavements in northern England. The total area of limestone pavement in England is 2340 ha, comprising 80% of the UK resource. Limestone pavement is listed as a priority habitat type in the EU Habitats Directive. England holds a large proportion of this habitat within Europe, with the only other significant areas being in Ireland and Sweden.

Limestone pavements support unusual combinations of plants. Plants grow mainly within the grikes, which provide sheltered, humid conditions with very thin soils. The grikes contain woodland and wood-edge plants such as herb Robert and dog's mercury, a number of ferns including hart's tongue, wall rue, male fern and the rare rigid buckler fern, which is almost exclusively confined to limestone pavement.

Trends and threats

Limestone pavement has been in demand as 'water-worn limestone' for garden rockeries since around 1870, but losses have accelerated over the last 40 years as extraction has become mechanised. Limestone rock has been removed both legally (under existing planning permissions) and illegally for garden use. A comprehensive survey of limestone pavements in 1975 estimated that, while 61% of the total area of limestone pavement in the UK was intact, only 3% was undamaged.

Overgrazing has also resulted in habitat degradation. In 1999, 40% of all pavements were in an unfavourable condition due to overgrazing, with a further 8% adversely affected by conifer plantations. In the longer term, there may be additional threats from climate change, including increased precipitation and acidification.

David Holmes

Figure 75 Plants grow in the crevices between limestone blocks

Managing this delicate ecosystem

Management is often focused on the improvement of habitat for plants and animals. This involves:

- ensuring that pavement is not extracted or damaged
- avoiding fertiliser application on or around pavement
- avoiding the use of 'asulam' herbicide for bracken control in the pavement area, as this will destroy ferns in grikes
- grazing with appropriate animals and suitable stocking levels, according to the type of pavement
- controlling rabbit populations

Additional management responses include better control over access and recreation, increased sustainability of agricultural practices and tighter legislation at both regional and national levels.

Question

(a) Examine why mountain ecosystems have low productivity.

(b) Design a programme of fieldwork to assess the impact on a new limestone cave system that has been opened recently as a tourist attraction.

Guidance

(a) They have low productivity in comparison with other ecosystems, for example rainforests and coral reefs. The main controls on this are climatic — low annual temperatures reduce the growing season and limit plant growth.

(b) In order to monitor impact, base line survey data are needed, i.e. on what the area was like prior to the new attraction. This will probably be based on a comparison survey in a similar habitat/area (geographically close). Some ideas might include:

- ecological surveys of plant diversity, bird populations etc.

- traffic and visitor surveys (including sphere of influence)
- using a decibel meter to assess noise

For the area that now hosts the attraction:

- repeat the above surveys (using an identical sampling strategy)
- carry out interviews with local shopkeepers, residents and other interested parties, and assess the reasons for their views
- develop an impact matrix for activities

Assess likely sources of secondary data — you could also use cost-benefit analysis.

Polar ecosystems

The polar regions are the most remote places on Earth, yet their conditions of extreme cold, altitude and wind, and because they are largely removed from the public eye and political priority, heighten their vulnerability. Figure 76 shows the location of the polar ecosystems.

Figure 76
Location of the south and north polar ecosystems

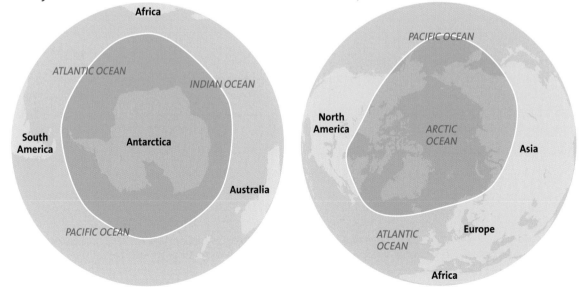

Differences between the polar ecosystems

The areas surrounding the two poles have common features of a cold climate, snow and ice. Otherwise, their land and marine ecosystems are significantly different.

A thick ice sheet covers the Antarctic continent; even during the summer season, only a few mountain and coastal areas are snow-free. The size of the ice sheet ranges from 4 million km^2 to 19 million km^2, depending on the season. It is, on average, 2.3 km thick, and represents 91% of the world's ice and the majority of the world's freshwater.

The Arctic, in contrast, consists of a large, deep ocean covered by drifting ice sheets a few meters thick. The land areas that surround the ocean, and are usually considered part of the Arctic region, are dominated by polar desert and tundra vegetation, although they include some prominent ice caps, such as Greenland's inland ice. The Arctic tundra is home to about 3.5 million people, many of whom make a living from marine and freshwater fishing, hunting and reindeer husbandry.

The importance of polar ecosystems

The polar regions offer a number of essential goods and services.

Regulation of global climate, ocean currents and sea level

The Earth's vast polar ice sheets serve as a mirror, reflecting a large percentage of the sun's heat back into space, thus keeping the planet cool. Without the ice sheets, additional heat from the sun would be retained in the ocean and more would be released into the atmosphere, feeding the warming process. Layers of peat and permafrost store vast amounts of carbon.

Biodiversity

Hundreds of species are endemic to the Arctic, a place where organisms have adapted to the extremes of temperature, daylight, snow and ice found in polar regions. The Arctic also serves as habitat for several migratory bird species and New Zealand's southern islands are home to around 250 species, including 35 endemics.

Food production

The Arctic marine waters are among the richest fishing regions in the world and a major contributor to the world's fish catches. In much of Newfoundland, Greenland, Iceland, the Faroe Islands and northern Norway, fishing is the primary industry. Secondary food sources may include moose, bighorn sheep, alpine hare, ducks, geese and other birds. Several polar fish stocks have been adversely affected in recent years. In the Faroe Islands, for example, cod landings decreased from about 200 000 tonnes to less than 70 000 tonnes between 1987 and 1993.

Recreation

There is a growing desire to explore polar areas. In the early 1990s, more than a million tourists were drawn to the Arctic. About 10 000 people visited Antarctica in 1998/9. Tourists can frighten wildlife, such as breeding penguins in Antarctica, leave behind rubbish, and create noise and pollution.

Feedback

The poles are important as early indicators of the pressures we are placing on global resources. For example, we can use analyses of the condition of the Arctic to better understand stratospheric ozone production, atmospheric cleansing and pollution transport in northern latitudes.

Although polar regions include some of the last large areas where human activity has not overtly altered the landscape, scientists have found solid evidence that human activities, often occurring in other parts of the world, are modifying polar environments and the goods and services they provide.

The diversity of polar ecosystems

Major controls on ecosystem types include weather patterns and ocean currents. Figure 77 illustrates the diversity of polar ecosystems.

Figure 77
The range of polar ecosystem types

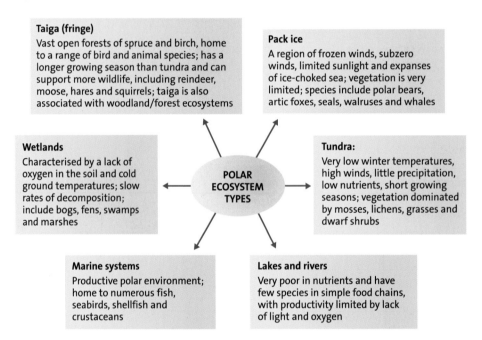

Taiga (fringe)
Vast open forests of spruce and birch, home to a range of bird and animal species; has a longer growing season than tundra and can support more wildlife, including reindeer, moose, hares and squirrels; taiga is also associated with woodland/forest ecosystems

Pack ice
A region of frozen winds, subzero winds, limited sunlight and expanses of ice-choked sea; vegetation is very limited; species include polar bears, artic foxes, seals, walruses and whales

Wetlands
Characterised by a lack of oxygen in the soil and cold ground temperatures; slow rates of decomposition; include bogs, fens, swamps and marshes

POLAR ECOSYSTEM TYPES

Tundra:
Very low winter temperatures, high winds, little precipitation, low nutrients, short growing seasons; vegetation dominated by mosses, lichens, grasses and dwarf shrubs

Marine systems
Productive polar environment; home to numerous fish, seabirds, shellfish and crustaceans

Lakes and rivers
Very poor in nutrients and have few species in simple food chains, with productivity limited by lack of light and oxygen

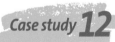

Case study **12** CLIMATE CHANGE AND ARCTIC WILDLIFE

Change is underway

The Arctic faces drastic change. The global build-up of greenhouse gases has the potential to alter the Arctic environment and its ecosystems dramatically. The Arctic is particularly vulnerable to climate change. Future climate change is likely to threaten both marine and terrestrial wildlife. From plankton to polar bears, many species could suffer or disappear entirely.

Evidence suggests the change is already well underway. A few of the recent signs of change are:

■ large fluctuations in wildlife populations
■ thinning sea ice over once-stable areas (Figure 78)

Contemporary Case Studies

- rising lake temperatures
- thawing permafrost
- regional warming

Annual temperatures have increased by about 4.5°C in areas such as Canada's Mackenzie River Basin, the Bering Strait and Lake Baikal in Siberia. This is many times the average global rate.

Inuit communities have documented these signs of change throughout the region. The Gwich'in Athabascan people in Arctic Village, Alaska, have observed marked changes in weather, hydrology, vegetation and animal distribution patterns, indicating that warming trends have occurred during the lifetimes of community residents. Scientific observations appear to confirm these indigenous accounts.

Warmer temperatures may increase biological productivity in some parts of the marine ecosystem. Reductions in sea ice extent and algae could, in turn, lead to profound reductions in overall biological productivity in the Arctic seas. The entire ice-associated community could be threatened, including its dependent fish species, such as polar cod. Arctic and migratory whale species, such as the narwhal and beluga, grey and bowhead whales, would be affected, since they feed along the ice edge.

1979–1981 (September average)

September sea ice coverage 1979–2005

2003–2005 (September average)

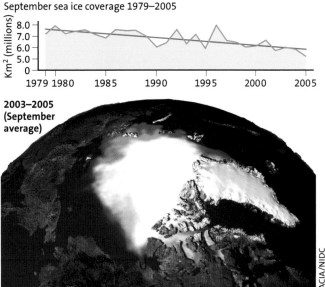

ACIA/NIDC

Figure 78 The reduction in area of the Arctic sea ice, 1979–2005

Rapid temperature increases in the Arctic are expected to push climate and vegetation zones northward at an astounding rate of 150 miles over 100 years. Many plant and animal species are unlikely to have time either to adapt to this warming, or to adjust their ranges to keep pace with the shift in climatic zones. Shifts in vegetation could jeopardise many animal populations. Grazing animals such as reindeer, caribou and musk ox closely track seasonal plant growth, and depend upon the vegetation for healthy herds and well-nourished calves. Without adequate vegetation, the animals become less productive, delay having offspring and, in many cases, starve.

The polar ecosystems are still relatively unmodified when compared with other ecosystems, but their once-pristine condition already shows signs of climate change and other pressures. Many biologists doubt that Arctic animals will be able to adapt to climate change. Arctic ecosystems are already considered 'stressed', or inherently vulnerable, compared with temperate and tropical ecosystems. This vulnerability is largely due to violent Arctic weather, which can destroy entire animal populations or generations of young birds. It is also due to the short and very sensitive nature of the Arctic food chain.

Desert ecosystems

Typically, a desert is defined as an area having an average precipitation of less than 25 cm per year. But a more important factor is that deserts have a strong tendency to lose water by evaporation, and this potential for water loss exceeds the annual rainfall. Deserts are characterised by having sparse vegetation, and the organisms that live in deserts are specially adapted to withstand or avoid water stress. Figure 79 shows the distribution of the world's major deserts. Deserts occupy nearly one-seventh of the world's land surface area, but carry only 4% of the human population.

Deserts tend to occur in two belts that circle the globe. Both the northern and southern hemispheres have this belt located between latitude 15° and 35°, roughly centred over the Tropics of Cancer and Capricorn. These areas receive intense sunlight as the sun is mostly overhead, and this solar energy heats the air. Hot and moist tropical air rises into the atmosphere. As this air rises, it cools, condensing the moisture and converting it to water that then falls as rain. Tropical air typically falls at about latitude 30° on either side of the equator, so the desert belt at the Tropics is hot and dry. The results are often persistent high pressure systems that tend to block incoming storms, or push them into other regions.

Desert classification

Desert biomes can be classified according to several characteristics.

Hot and dry desert

Figure 79 *The deserts of the world*

These include deserts in North America, southern Asian, South and Central America, Ethiopia and Australia. The seasons are generally warm throughout the year and

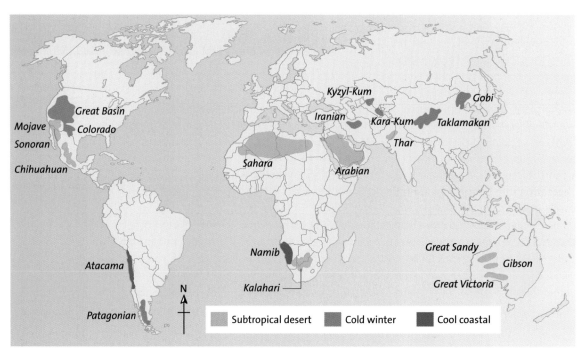

very hot in the summer. Mean annual temperatures range from 20°C–25°C. Minimum temperatures sometimes drop to –18°C. Rainfall is usually very low and/or concentrated in short bursts between long rainless periods.

Semi-arid desert

These deserts are situated in parts of North America, Newfoundland, Greenland, Russia, Europe and northern Asia. The summers are moderately long and dry; the winters normally bring low concentrations of rainfall. Summer temperatures usually average around 21–27°C. Cool nights help both plants and animals by reducing moisture loss from transpiration, sweating and breathing. As in the hot desert, rainfall is often very low and/or concentrated. The average rainfall range is 20–40 mm yr^{-1}.

Coastal desert

These deserts occur in moderately cool to warm areas — a good example is the Atacama of Chile. The cool winters of coastal deserts are followed by moderately long, warm summers. The average summer temperature ranges are 13–24°C and winter temperatures are 5°C or below. The average rainfall is around 8–13 cm.

Cold desert

These deserts have cold winters with snowfall and high overall rainfall throughout the winter. They occur in the Antarctic, Greenland and the Nearctic Realm (North America, Canada, Alaska and Greenland). The mean annual precipitation range is 15–26 cm. The mean winter temperature is between –2°C and +4°C and the mean summer temperature is 21–26°C.

Biodiversity

Figure 80
Saguaro cactus

Surprisingly, there are many species of plant that survive in the desert. Most of these are succulents, which means they store water. Others have seeds that lay dormant in the sand until rain falls. Somehow, these plants find a way to absorb water and protect themselves from the heat. Adaptation to the harsh environment is central to survival.

The most famous desert plant type is the cactus, of which there are many species. The saguaro is the familiar tall, pole-shaped cactus (Figure 80). It can grow up to 12 m tall and can hold several tonnes of water inside its soft tissue. Like all cacti, the saguaro has a thick, waxy layer that protects it from the sun. Other succulents include the desert rose and the living rock, which is disguised to protect it from being eaten. *Welwitschia* is a peculiar looking plant. It has two long leaves and a big root. This plant is actually a type of tree and it can live for thousands of years.

One of the most common and destructive desert pests is the locust — a relative of the grasshopper. Locusts travel from place to place, eating all the plants they find. They can destroy many crops in a single day. There are also arachnids in the desert, including spiders and scorpions.

A sensitive ecosystem

The Great Basin Desert covers a large area of land, bounded by the Sierra Nevada Cascade mountain range to the west, and the Rocky Mountain range to the east (Figure 81). Between these large mountain ranges is a series of valley floors (basins) at an elevation of about 1200 m. This is a cool or cold desert due to its more northern latitude, as well as higher elevations.

Today the Great Basin Desert is a region of wild beauty and extreme ecological fragility. There are a number of threats to this precious environment:

- invasive species are threatening local endemic populations
- livestock, grazing on private and public land, often trample stream sides, damaging vegetation
- cyanide from gold mines has contaminated water sources
- intensification of agriculture means that irrigation flushes boron, arsenic and mercury out of the soil into streams and ponds, poisoning of fish and wildlife
- there is illegal dumping of toxic wastes in remote areas
- wildfires occur regularly
- urban pressure on the desert's periphery drains off water, threatening to turn an already arid region into a genuine wasteland

There are a range of management strategies in place to try to preserve these special ecosystems. The Great Basin has become a National Park, which affords it some conservation status and protection. Management strategies need to be coupled with habitat inventories to monitor changes and losses.

Figure 81
The Great Basin Desert ecosystem

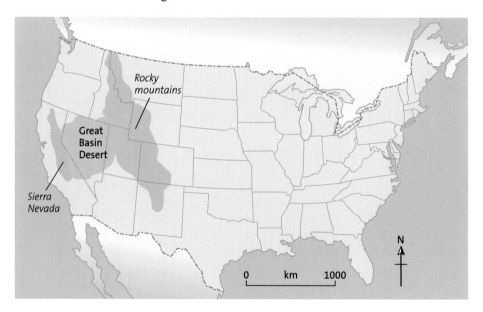

Desertification

Desertification is the degradation of land in arid, semi-arid and dry sub-humid areas resulting from various factors, including climatic variations and human activities.

Modern desertification often arises from the demands of increased populations that settle on the land in order to grow crops and graze animals.

Desertification is widespread in many areas of China. The populations of rural areas have increased since 1949 for political reasons as more people have settled there. While there has been an increase in livestock, the land available for grazing has decreased. The import of European cattle such as Friesian and Simmental, which have higher food intakes, has also made the situation worse.

Human overpopulation is leading to the destruction of tropical wet forests and tropical dry forests, due to widening practices of slash-and-burn and other methods of subsistence farming necessitated by famines in less developed countries. Large-scale erosion and loss of soil nutrients often follow deforestation and sometimes lead to desertification. Examples of this extreme outcome can be seen on Madagascar's central highland plateau, where about 7% of the country's total land mass has become barren, sterile land.

Summary

1 Ecosystems of low productivity include a range of contrasting habitat types, but they are all linked by the fact that they are limited by climatic extremes, for example, too cold, too wet or too dry.
2 Low productivity systems are generally found in areas of low population density; this is in part a reflection of their status and of the harsh environment.
3 Despite their low productivity, many of these ecosystem types have a surprisingly high biodiversity and are nationally or internationally important.
4 There are a range of threats to low productivity systems. In particular, increasing tourism and encroaching urban areas are using up valuable resources within an already stressed environment.
5 The future of such areas has a mixed outlook. In general, there has been a decline in the spatial area and quality of low productivity ecosystems. However, scientists and policymakers are beginning to recognise that such areas have an important role to play, not only in terms of sustaining local indigenous populations, but also contributing to the earth–atmosphere climate regulation system.

Websites

http://www.biodiv.org/programmes/areas/mountain/default.asp
Information about mountain ecosystems.

http://pubs.usgs.gov/gip/deserts/desertification/
USGS publication about desertification.

http://www.gluckman.com/ChinaDesert.html *An article about desertification in China.*

Human ecosystems

The urban ecosystem

Urban areas occupy only around 2% of the land surface area of the planet, but provide a home for half of the world's population — about 3 billion people. The proportion of people living in cities is even higher in the developed regions of the world. In the US and UK, for instance, more than three-quarters of the population live in urban areas.

Profile of urban ecosystems

Urban ecosystems are perhaps the most familiar of all ecosystems. They contain a rich patchwork of green spaces — parks, yards, street plantings, greenways, urban streams, commercial landscaping and un-built plots — that provide the living heart of the urban ecosystem (Figure 82).

Figure 82 Sheffield Metropolitan District, containing part of the Peak District National Park

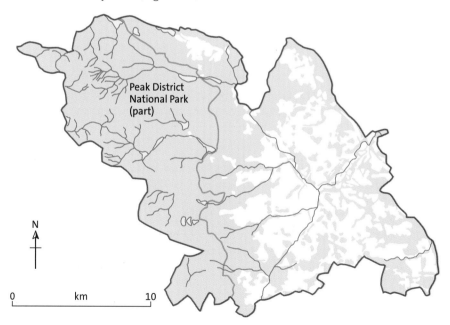

Peak District
National Park
(part)

N

0   km   10

There are many differences between urban ecosystems and those less influenced by humans.

- Urban ecosystems are generally highly disturbed, and subject to rapid changes in soil and plant cover, as well as temperature and water availability.

- Plants are characteristically non-native garden plantings and weeds. Even in the city's natural or semi-natural areas, such as parks, the vegetation is often greatly altered, with many non-native and invasive species.
- Environmental stresses also modify the natural elements of urban ecosystems. Trees, for instance, are subject to high levels of air pollutants, road salt and runoff, physical barriers to root growth, disease, poor soil quality, frequent drought and reduced sunlight.

Importance of the urban ecosystem

Urban green spaces (Figure 83) contribute their own unique and essential services to the urban system.

Shade and temperature control

Street trees and other green spaces, help to reduce the urban heat-island effect in the summer. Temperatures in the heart of a city may be 0.6–1.3°C warmer than in rural areas, due to the large heat-absorbing surfaces of buildings and asphalt, as well as high energy use.

Air filtering

City plants are efficient air pollution removers. In a park, the leaf surfaces of trees can filter out as much as 85% of the ambient air pollution — mostly particulates. Street trees can also be effective air purifiers, removing up to 70% of particulates on a tree-lined street.

Noise reduction

Trees and shrubs can help filter out noise pollution too. A 30 m belt of tall dense trees, combined with soft surfaces, can reduce local noise levels by 50%.

Runoff control

Urban forests, wetlands and streamside vegetation help to restore some of the natural balance by buffering storm water runoff, absorbing pollutants and

Figure 83
Urban green spaces provide important wildlife sites

recharging groundwater reservoirs. This may help to reduce the 'flashiness' of the hydrographic response and reduce the risk of flooding.

Biodiversity and wildlife habitat

Cities support a variety of plants and animals. In fact, the extent of 'urban wildlife' is often surprising. Urban ecosystems also provide a critical habitat to many migratory species. For example, urban green spaces and parks are critical to many migratory birds that otherwise would face long stretches with no resting and feeding places.

Recreation, aesthetic, and spiritual values

Parks and green spaces provide city dwellers with invaluable recreational opportunities, such as biking, jogging, skating and dog-walking.

Food production

Worldwide, some 800 million city residents grow food in backyards, vacant lots, roadsides and small suburban farms. In Kenya and Tanzania, two out of three urban families are engaged in farming. In Taiwan, more than half of all urban families are members of farming associations. In Cuba in 2003, urban agriculture produced 800 000 tonnes of fresh organic produce and employed 165 000 people.

Figure 84 illustrates the range of green spaces within a city.

Figure 84
Range of green spaces within a city

Formally designated open spaces

Woods — Burial places

Paved city spaces with plants — Parks, gardens and sports grounds

A CITY'S GREEN SPACES

Farmland and horticulture — Domestic gardens

Water margins — Water — Private open spaces

Allotments — Transport corridor verges — Derelict land and tips

Other actual green spaces

Case study 14 — LONDON'S WASTELANDS

The unofficial countryside

There has been a substantial reduction in the extent of London's urban wasteland since the mid-1990s. For example, the former Docklands area contained a significant proportion of the capital's urban wastelands, but most of this area has been redeveloped to accommodate London's burgeoning service sector industries. Other large areas have been lost in more recent years to provide land for new housing. Despite the losses of some of the most extensive areas of urban wasteland in London, new sites, albeit smaller and more widely dispersed, are continually being created as a result of abandonment — a feature of the development cycle in a major conurbation.

Nature conservation importance

On the whole, wastelands offer comparatively high levels of biodiversity. They encompass a wide range of sites with varying substrates, topographies and other factors that determine the distribution of plant and animal species.

A common feature of many urban wastelands is the dominance of species that are considered to be 'weedy' pioneer species. These species are best able to colonise disturbed or hostile environments, but often succumb to competition once conditions stabilise. For this reason, many of the species that flourish in urban wastelands are exotics that would normally be out-competed by native species, or are species that have

Figure 85 *An example of urban wasteland*

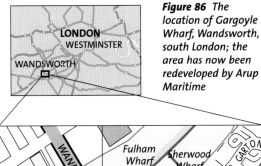

Figure 86 *The location of Gargoyle Wharf, Wandsworth, south London; the area has now been redeveloped by Arup Maritime*

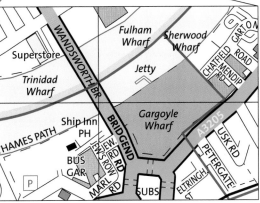

exacting climatic or biological requirements, which are rarely available except in the unusual conditions that arise on urban wastelands.

There are a number of plants and animals that are characteristic of London's wastelands (Table 22).

Gargoyle Wharf is a classic wasteland site next to the River Thames in the London Borough of Wandsworth, and was once a proposed Site of Metropolitan Importance for Nature Conservation (Figure 86). It was the subject of at least three major planning applications over a period of 6 years and there was a vigorous local campaign for the

Table 22 *Flora and fauna characteristic of wasteland in London*

Species	Name	Comments
Black redstart	*Phoenicurus ochruros*	A red-tailed, robin-sized bird of the thrush family and one of Britain's rarest birds; once known as the 'power station' or 'bombsite' bird; it is associated with wasteland and industrial sites
Linnet	*Carduelis cannabina*	Wasteland with areas of scrub and grassland provide essential feeding and breeding grounds for these birds throughout the year; males in summer plumage sport striking splashes of carmine red
Common lizard	*Lacerta vivipara*	These reptiles like quiet sites with open ground that retain heat — conditions often found on wasteland sites
The humble bumble bee	*Bombus humilis*	The brown-banded Carder bee is another name for this nationally scarce bee; one of the national strongholds for this species is the flower-rich wasteland habitat found in the Thames Gateway area of London
False London rocket	*Sisymbrium loeselii*	This was introduced to London, and is found widely scattered across London's wasteland sites
Rosebay willowherb	*Chamerion angustifolium*	'Fireweed' is a classic plant of disturbed ground; historically, it is associated with the bombsites of the Second World War and springs up in areas where there have been fires
Teasel	*Dipsacus fullonum*	Teasel is often found in the rough grassland of some wasteland sites; its name derives from the use of its spiny heads to tease wool before spinning; goldfinches are often seen feeding on the seedheads in winter

open space to be retained. Over 250 different species of plants were recorded on the site, before it was bulldozed ready for development in 2002.

Wastelands are under threat as they are mostly brownfield sites. It is sometimes unfortunate that such areas are the focus for redevelopment and regeneration, to avoid building on greenfield locations.

Question

Why is it especially important to protect urban wildlife corridors (i.e. canals, railways, hedgerows and roadside verges)?

Guidance

Corridors provide a means of movement or dispersal for a variety of animal and plant populations. Corridors link isolated fragments of habitat together. These artificial structures are critical for species with narrow habitat ranges and a limited ability to disperse over large areas. Examples of animals that rely on corridors to find new areas and expand populations include wood mice and toads. Sometimes corridors are exploited by exotic plants and animals, e.g. Himalayan balsam, Japanese knotweed and mink.

Agro-ecosystems

'…the farmers on Salisbury Plain have transformed this landscape on a scale that would have seemed unthinkable even a generation ago…across the length and breadth of Britain the countryside has been reconstructed in the sole interests of intensive agriculture.'

Graham Harvey (agricultural story editor for *The Archers*), *Killing the Countryside* (1997)

Throughout history, including much of the twentieth century, agricultural output has been increased mainly through bringing more land into production (see Table 23). This has been combined with agricultural practices such as seed beds, improved animal nutrition and water management. Societies have flourished to a great degree by improving their capacity to expand agricultural production.

The world's agro-ecosystems provide the overwhelming majority of food and feed on which humanity depends for its continued well-being. In 1997, agriculture provided 94% of the protein and 99% of the calories consumed by humans (FAOSTAT, 1999).

Agro-ecosystems provide a range of goods and services.
Goods include:
- food crops
- fibre crops
- crop genetic resources

Services include:
- maintaining limited watershed functions (infiltration, flow control, partial soil protection)

- provision of habitat for birds, pollinators and soil organisms important to agriculture
- building of soil organic matter
- sequestration of atmospheric carbon
- provision of employment

Environmental concerns

The unprecedented scale of agricultural expansion and intensification (see Table 23) has raised two major environmental concerns.

First, there is growing concern over the vulnerability of the productive capacity of many agro-ecosystems to the stresses imposed on them by intensification. Local evidence points to such problems as soil salinisation caused by poorly managed irrigation systems, loss in soil fertility through over-cultivation of the fragile soils of the tropical savannas, and lowering of water tables through over-pumping of water for irrigation purposes. But the global significance of this degradation is still far from clear.

The second concern relates to the negative environmental impacts of agricultural production that intensification often amplifies. Such externalities have complex and sometimes far-reaching consequences. Examples include the negative impact of increased soil erosion from hillside farming on downstream fisheries and hydraulic

Table 23
Agricultural extent and type by region, 2003

Region	Total land area (000 m²)	Total agricultural Area (000 m²)	Total agricultural Share of total land area (%)	Crop land Area (000 m²)	Annual crops Area (000 m²)	Annual crops Harvested area (000 m²)	Permanent crops Area (000 m²)	Permanent crops Harvested area (000 m²)	Permanent pasture Area (000 m²)	Irrigated land Area (000 m²)
North America	18 722	4 991	26.7	2 306	2 284	1 223	22	17	2 685	221
Latin America and the Caribbean	20 178	7 552	37.4	1 531	1 279	924	251	154	6 004	171
Europe	4 726	2 156	45.6	1 365	1 227	837	138	130	791	167
Former Soviet Union	17 877	5 863	32.8	2 299	2 246	1 288	53	35	3 555	204
West Asia/ North Africa	11 890	3 682	31.0	920	816	571	104	92	2 762	239
Sub-Saharan Africa	22 676	9 909	43.7	1 659	1 465	1 207	193	156	8 254	62
Asia	24 703	10 304	41.7	4 443	3 975	4 168	468	380	5 861	1 434
East Asia	16 213	6 995	43.1	1 503	1 407	1 506	95	100	5 492	580
South Asia	4 129	2 231	54.0	2 035	1 946	2 023	89	106	196	712
Southeast Asia	4 360	1 078	24.7	905	621	640	284	174	173	142
Oceania	8 491	4 826	56.8	518	492	167	25	11	4 308	24
Global total	130 484	49 281	37.8	15 039	13 785	10 386	1 254	975	34 219	2 523

infrastructure, and the damage to both aquatic ecosystems and human health arising from fertiliser and pesticide residues in water sources or on crops.

There are also concerns about the loss of habitat and biodiversity from converting land to agricultural use, as well as narrowing the genetic base and genetic diversity of domesticated plant and animal species currently in use. Globally, scientists recognise that agriculture influences climate by altering global carbon, nitrogen and hydrological cycles.

Figure 87 shows how different agricultural systems affect species diversity.

Figure 87 How different agricultural systems affect species diversity

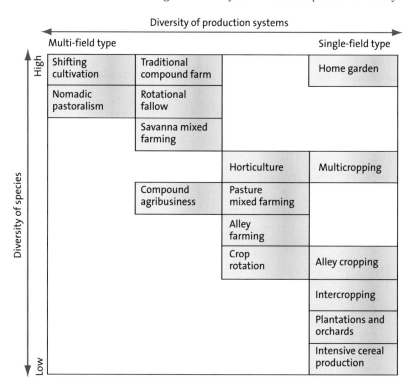

Intensification and the collapse of farmland bird populations

The RSPB (Royal Society for the Protection of Birds) has suggested that there is a link between agricultural intensification and the collapse of Europe's farmland bird populations. Figure 87 shows the relationship between agricultural intensity (cereal yield) and the trend in farmland bird population in European countries.

The data in Figure 88 highlight the fact that farmland bird populations declined in most European countries during the period 1970–90. It also shows that there is a considerable variation between countries. These differences between countries can be explained in terms of agricultural intensity. Population declines were greater in European Union (EU) countries, subject to the Common Agricultural Policy (CAP), than in former Communist countries. The UK has one of the greatest cereal yields and, perhaps unsurprisingly, the steepest decline in farmland bird populations. These results show that agricultural intensification can affect bird populations on a continental scale and should be regarded as a major threat to biodiversity.

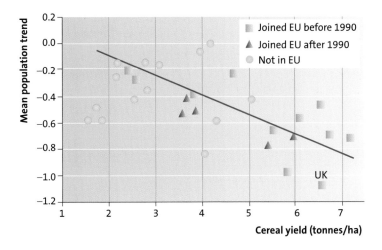

A case of successful sustainable agriculture

The Cuban experience illustrates that it is possible to feed a nation's population using small- or medium-sized farms based on appropriate ecological technology, and in doing so can allow the nation to become more self-reliant in food production. Farmers receive higher returns for their produce, and when they do, they are encouraged to produce. Capital-intensive chemical inputs (most of which are unnecessary) have been largely dispensed with.

Key elements of sustainability

- Agro-ecological technology instead of chemicals: Cuba has used intercropping, locally produced bio-pesticides, compost and other alternatives to synthetic pesticides and fertilisers.
- Fair prices for farmers: Cuban farmers stepped up production in response to higher crop prices. Farmers in other parts of the world lack an incentive to produce when prices are kept artificially low, as they often are. Yet, when given an incentive, they produce, regardless of the conditions under which that production must take place.
- Redistribution of land: small farmers and gardeners have been the most productive of Cuban producers under low-input conditions. Indeed, smaller farms worldwide produce much more per unit area than do large farms.
- Greater emphasis on local production: locally and regionally produced food offers greater security, as well as synergistic linkages to promote local economic development. Furthermore, such production is more ecologically sound, as the energy spent on international transport is wasteful and environmentally unsustainable. By promoting urban farming, cities and their surrounding areas can be made virtually self-sufficient in perishable foods, be beautified and have greater employment opportunities. Cuba gives us a hint of the under-exploited potential of urban farming.

Figure 89 Location of Cuba

Question

Figure 90 shows the value of production per agricultural worker (1995–97 average) by region. Examine why farming systems, agricultural technology and the mix of production inputs vary markedly across regions.

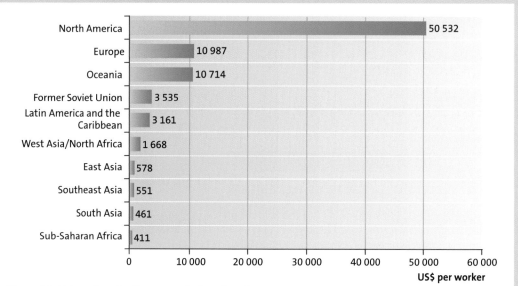

Figure 90 *Value of production per agricultural worker, 1995–97 average*

Guidance

Capital-intensive systems are evident in North America, Europe and Oceania, whereas Africa and Asia have a much lower output per head of population. The availability of land, labour, technology and capital resources (and, hence, their relative prices) affect the variation directly. There is a complex association between factors here. In MEDCs, technology and capital are available for agricultural intensification. Advanced mechanisation has taken place, partly due to the relatively high costs of labour and production. Crops are grown on a commercial basis. In LEDCs, the opposite may be the case. In small-scale family-run businesses (or cooperatives), the farming operation is labour intensive and uses few outside inputs.

Summary

1 There are a variety of human-influenced ecosystems, which include urban and agricultural systems. Their common feature is that they are either managed or controlled directly by people.

2 Urban ecosystems may be undervalued, but really represent 'unofficial' countryside. They are the essential green or brown 'lungs' of metropolitan areas. Even brownfield sites, with their associated image of wastelands, are important habitats for a range of plants and animals.

3 Transport infrastructure and routeways also act as important corridors for populations to migrate along and help prevent genetic isolation.

4 Intensification of agriculture has been significant in many developed countries since 1945. This has brought consumer benefits in the form of improved food security and lower prices, but has also had a significant impact on farm wildlife, especially bird populations.

Websites

http://www.pbs.org/earthonedge/ecosystems/urban.html
A good article on urban ecosystems.

http://www.ifpri.org/pubs/abstract/130/rr130toc.pdf#search=%22AGRICULTURAL%20INTENSIFICATION%20BRAZIL%20REPORT%22
An interesting article on impacts of agricultural intensification in Brazil.

http://www.arkiveeducation.org/urban_habitat_notes.pdf#search=%22urban%20ecosystems%22
The range of urban habitats is discussed.

Biodiversity: loss and conservation

Biodiversity can be defined as follows:

> The variety among living organisms from all sources including, amongst others, terrestrial, marine and other aquatic ecosystems and the ecological complexes of which they are part; this includes diversity within species, between species and of ecosystems.

The Convention on Biological Diversity

Biodiversity is simply the variety of life on earth. The term is used to summarise the phrase 'biological diversity', which includes:

- species diversity
- genetic diversity
- ecosystem and habitat diversity

Biodiversity is not just about rare, unusual or endangered organisms; it also concerns the plants and animals familiar to all of us in the places where we live and work.

Biodiversity concerns

Global stocks of biological diversity generate a flow of goods and services to all societies on Earth (Figure 91).

*Figure 91
Why biodiversity is
important*

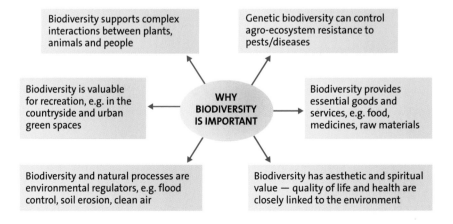

Scientists are concerned about the rates of biodiversity loss that results from human activity. The UK, for example, is estimated to have lost over 100 species in the last century. However, this figure should be taken in context. The data in Table 24 estimate the total number of global species at around 12.5 million, although some estimate this figure to be nearer 80 million.

A total of 15 589 species (7266 animal species and 8323 plant and lichen species) are now considered at risk of extinction. Current extinction rates are at least one hundred to a thousand times higher than background, or 'natural' rates.

IUCN Red List of Threatened Species

Groups	Known number of species	Estimated total number of species	% known species
Insects	950 000	8 000 000	11.9
Fungi	70 000	1 000 000	7.0
Arachnids	75 000	750 000	10.0
Nematodes	15 000	500 000	3.0
Viruses	5 000	500 000	1.0
Bacteria	4 000	400 000	1.0
Plants	250 000	300 000	83.3
Protozoans	40 000	200 000	10.0
Algae	40 000	200 000	10.0
Molluscs	70 000	200 000	35.0
Crustaceans	40 000	150 000	26.7
Vertebrates	45 000	50 000	90.0
World total	1 700 000	12 500 000	13.6

Table 24 *Known and estimated numbers of species on Earth*

Factors influencing biodiversity

While there are many concerns regarding biodiversity loss, there are also positive factors that increase biodiversity, as well as negative ones that threaten it (Table 25).

	Factors encouraging biodiversity	Factors threatening/reducing biodiversity
Natural factors	■ High temperature, light and humidity levels, together with a lack of seasonality encourage high primary productivity ■ A greater range of altitudes provides a more diverse range of habitats	■ Limiting factors to growth, e.g. cold/dry environment (reduces both range and total numbers of species) ■ Small population size and isolation increases vulnerability to disaster
Human factors	■ Large (undisturbed) continuous biomes/eco-regions provide a greater range of ecological niches ■ Well managed habitats coupled with 'bottom-up' conservation strategies, including effective ecotourism	■ Over-hunting/over-fishing/exploitation ■ Population growth

Table 25
Factors influencing biodiversity

A more complex picture of human biodiversity threats is presented in Figure 92.

Ecosystems & Biodiversity

Figure 92 *Threats to biodiversity*

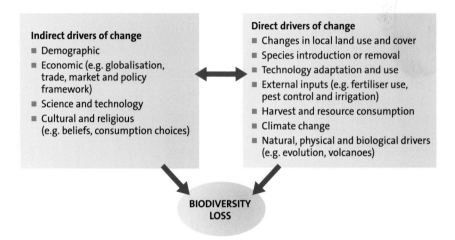

Indirect drivers of change
- Demographic
- Economic (e.g. globalisation, trade, market and policy framework)
- Science and technology
- Cultural and religious (e.g. beliefs, consumption choices)

Direct drivers of change
- Changes in local land use and cover
- Species introduction or removal
- Technology adaptation and use
- External inputs (e.g. fertiliser use, pest control and irrigation)
- Harvest and resource consumption
- Climate change
- Natural, physical and biological drivers (e.g. evolution, volcanoes)

BIODIVERSITY LOSS

Changes in drivers that affect biodiversity *indirectly*, such as population, technology and lifestyle, can lead to changes in drivers affecting biodiversity *directly*, such as the catching of fish or the application of fertilisers to increase food production. These result in changes to biodiversity and to ecosystem services, thereby affecting human well-being. These interactions can take place at more than one scale and can cross scales. For example, international demand for timber may lead to a regional loss of forest cover, which increases flood magnitude along a local stretch of a river. Similarly, the interactions can take place across different time scales, i.e. short, medium or long term.

Adapted from *Millennium Ecosystem Assessment*, World Resources Institute (2005)

Global distribution of biodiversity

Biodiversity varies at local, regional, national and international levels. Figure 93 illustrates the distribution of some of the most highly valued terrestrial biodiversity

Figure 93 *Global distribution of terrestial biodiversity, where red = high biodiversity and blue = low biodiversity*

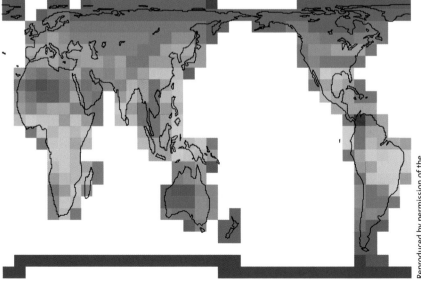

worldwide (mammals, reptiles, amphibians and seed plants). Red indicates high biodiversity and blue indicates low biodiversity. Note how there is considerable variation in biodiversity with latitude.

Table 26 shows the world ranking of mega-biodiversity countries.

Most of the mega-biodiversity countries are LEDCs. These countries have the least amount of money to develop sophisticated conservation strategies in order to manage the underlying causes of biodiversity loss. Closure (or 'extreme protection') of such ecoregions can threaten the existence of people who are often living below the poverty line.

Country	Mammals	Birds	Flowering plants
Brazil	394	1 635	55 000
Colombia	359	1 695	50 000
China	394	1 244	30 000
Indonesia	436	1 531	27 500
Mexico	450	1 026	25 000
South Africa	247	790	23 000
Venezuela	305	1 296	20 000
Ecuador	302	1 559	18 250
Peru	344	1 678	17 121
India	316	1 219	15 000
Malaysia	286	736	15 000
Australia	252	751	15 000
Zaire	415	1 096	11 000
Costa Rica	205	850	11 000
Papua New Guinea	214	708	10 000
Panama	218	926	9 000
Madagascar	105	253	9 000
Cameroon	297	874	8 000
Philippines	153	556	8 000
Vietnam	213	761	7 000

Table 26 World ranking of mega-biodiversity countries

15

Using case studies

Question
(a) Describe and account for the global distribution of biodiversity.
(b) Why could the location of biodiversity hot spots in LEDCs create problems in terms of their management and conservation?

Biodiversity: the UK Action Plan

The UK government was one of the first signatories to the Convention on Biological Diversity (one of several major initiatives from the Rio Earth Summit in 1992) and established the UK Biodiversity Steering Group in 1994. The objectives for conserving biodiversity that underpin the UK Action Plan are to conserve and protect and, where practical, enhance:

- the overall population and natural ranges of native species
- internationally important and threatened species, habitats and ecosystems
- species, habitats, and natural and managed ecosystems that are characteristic of local areas
- the biodiversity of natural and semi-natural habitats, where this has been diminished over recent decades

In 1995, the Steering Group recommended the development of local Biodiversity Action Plans (BAPs). These represented an important new approach to nature conservation — the emphasis on partnerships working to commonly agreed goals. The functions of local biodiversity plans are as follows:

- to ensure that national targets for species and habitats are translated into effective action at the local level
- to identify targets for species and habitats appropriate to the local area, reflecting the values of people locally
- to develop effective local partnerships to ensure that programmes for biodiversity conservation are maintained in the long term
- to raise awareness of the need for biodiversity conservation in the local context
- to ensure that opportunities for conservation and enhancement of the whole biodiversity resource are fully considered
- to provide a basis for monitoring progress in biodiversity conservation at both local and national level.

BIODIVERSITY IN SHROPSHIRE

Saving wildlife

Shropshire is England's largest land-locked county (Figure 94) and has a wide diversity of habitats that reflect the physical characteristics of geology and landscape. Factors such as soil structure and chemistry are in turn influenced by slope, drainage and aspect to give many subtle variations in conditions. As a result of its central location within Britain, Shropshire is the meeting point for species of several different geographic ranges.

The richness of Shropshire's biodiversity is reflected in the number of statutory and non-statutory sites designated for nature conservation: four National Nature Reserves (NNRs), 111 Sites of Special Scientific Interest (SSSIs) and 573 non-statutory Wildlife Sites (formally known as Prime Sites for Nature Conservation). NNRs, SSSIs and Wildlife Sites together cover approximately 6% of the land area of Shropshire. In addition, the Shropshire Hills Area of Outstanding Natural Beauty (AONB) covers 23.2% of the land area.

Losing ground...

In the period 1980–1990, over 10% of Shropshire's best wildlife habitat was destroyed by agriculture (64%), forestry (21%) and urban development (15%). Figures 95 and 96 demonstrate the extent of the losses of woodland and flower-rich grassland in the county. The left-hand map in both of these figures shows the total area of each habitat in each 10 km by 10 km square. The circles are proportional to the total area of that habitat in the square, as shown in the key for Figure 95. The right-hand map in the two figures shows

Figure 94 *Shropshire — one of the largest land-locked counties in England*

Designation	Area (ha)	% of county
National Nature Reserve	1658	0.5
Sites of Special Scientific Interest (includes RAMSAR sites (protected wetlands) and SACs (Special Areas of Conservation)	8715	2.5
Wildlife sites	c. 10000	2.9
Area of Outstanding Natural Beauty	80916	23.2

Table 27 *Shropshire's conservation sites*

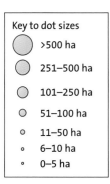

Key to dot sizes
- >500 ha
- 251–500 ha
- 101–250 ha
- 51–100 ha
- 11–50 ha
- 6–10 ha
- 0–5 ha

Area of hxabitat in each 10 km × 10 km square

Proportion lost (black), 1980–90

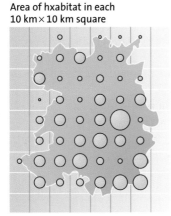

Figure 95 *Woodland in Shrophire — extent and loss*

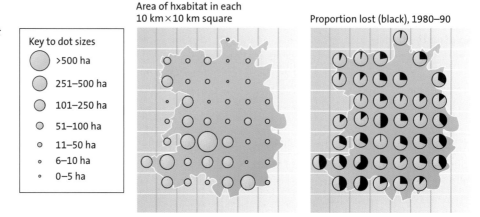

Figure 96 Flower-rich grassland in Shropshire — extent and loss

Key to dot sizes
- ⬤ >500 ha
- ● 251–500 ha
- ● 101–250 ha
- ○ 51–100 ha
- ○ 11–50 ha
- ○ 6–10 ha
- ○ 0–5 ha

Area of hxabitat in each 10 km × 10 km square

Proportion lost (black), 1980–90

the proportion of the original habitat lost from each 10 km by 10 km square. The dark portion of each circle is the proportion lost from the total, shown by the map on the left. If no circle is shown, then there has been no recorded loss from that square.

The Shropshire Biodiversity Action Plan (BAP)

Part 1 of the BAP describes the richness of biodiversity in the county, and explains the underlying factors that have contributed to that richness.

The plan sets out overall aims and key objectives and clarifies why particular habitats and species have been selected for individual action plans. Important issues, such as the collection and dissemination of information and data, raising awareness and involvement, and monitoring progress, are reviewed. Links are also made between the plan and other important initiatives, such as sustainable development and community strategies.

Table 28 Part of the Shropshire BAP

8 Action plan: urban areas 8.1 Policy and protection	Meets objective	Lead organisation	Assisting organisation	By when?
■ Ensure that all urban green spaces meeting SSSI or LNR criteria are designated	A	EN, SWT	LAs	2006
■ Identify all green spaces that meet Wildlife Site criteria and ensure their protection through the local planning process	A	SWT, LAs		2006
■ Include policies for the protection and enhancement of urban biodiversity in local plans and associated Supplementary Planning Guidance	A, C	LAs	EN, SWT	Ongoing
■ Encourage the production of an Urban Green Network guidance document to give advice on all aspects of Shropshire's urban green space	A, B, C	LAs	SWT, EN	2003

EN = English Nature; SWT = Shropshire Wildlife Trust; LAs = Local Authorities; LNR = Local Nature Reserve
Objective A Protect all sites of nature conservation importance in urban areas.
Objective B Increase the extent of green space in urban areas through restoration of degraded or impoverished land, or through habitat creation.
Objective C Maintain and enhance existing areas of urban green space through appropriate management techniques

Part 2 of the plan presents detailed action plans for 15 habitats and 33 species of priority concern in Shropshire. Basically, the action plan identifies what needs to be done for the threatened species and habitats of Shropshire, which organisation will lead the action for each, and the time-scale for the work.

Table 28 shows part of the urban section of the Shropshire BAP, including which organisation is responsible and the appropriate timescale. Table 29 shows another extract from the Shropshire BAP, giving reasons for habitat selection.

Habitat action plans	Reasons for selection
Semi-natural, broad-leaved woodland	Includes several woodland UK BAP Priority Habitats: upland oakwood, upland mixed ashwoods and wet woodland; also includes ancient and other areas of semi-natural woodland of local and regional significance
Veteran trees and parkland	Includes the UK BAP Priority Habitat lowland pasture and parkland; characteristic features of the Shropshire landscape
Hedgerows	Includes the UK BAP Priority Habitat ancient and/or species-rich hedgerows; many Shropshire hedgerows are very valuable for wildlife or have considerable potential for enhancement; they are important for connecting fragmented habitats

Table 29 Shropshire habitat and species action plans: reasons for selection

Conservation

We have already seen that there is a worrying decline in the health of many eco-systems, and in their quality and geographical extent. Maintaining the health of ecosystems is important for a number of reasons:

- Species diversity influences ecosystem stability — a variety of plant/animal species is needed to carry out ecosystem functions (e.g. cycling carbon) with maximum efficiency. Diversity bolsters an ecosystem's resilience, i.e. its ability to withstand climatic change, such as long-term drought hazard or the impact of global warming.
- Genetic diversity can determine an agro-ecosystem's resistance to pests and diseases, thus influencing its productivity — hence the concern about plant breeding, which tends to reduce naturally diverse polycultures to artificial mono-cultures (rice/wheat etc). The gene pool is fundamental to human health and the combating of disease. The loss of knowledge from indigenous communities of people under threat and the physical loss of key species, threatens traditional medicine and the search for potential drug ingredients.
- Ecosystem (habitat) diversity involves the complete food web, and the biotic and abiotic components that represent a visual resource for tourism (e.g. coral reefs and rainforests).

A need for conservation

A number of issues exist when considering how best to conserve biodiversity. Scientists and politicians need to think about the following:

- **methods** of conservation
- **what** to conserve
- **how** to do it

These are discussed more fully at the end of this section, on pages 111–112. Like many global issues, conservation requires a global framework for action, combined with the effective operation of national, regional and local strategies. There have been a number of changes in approaches to conservation (Figure 97).

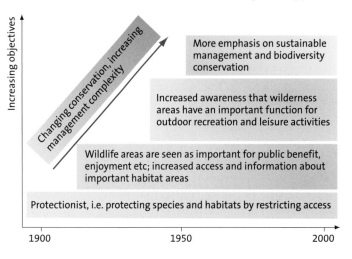

Figure 97 *Changes in approaches to conservation*

Up until the early 1990s, biodiversity loss was looked at as a scientific global issue, with a whole series of frameworks for species protection and habitat conservation.

- Measures for species protection included CITES (Convention on International Trade in Endangered Species) 1991, which banned trade in threatened species and their products (e.g. live parrots, ivory and reptile skins).
- Measures for habitat conservation included a combination of conventions such as the Convention on Wetlands, signed in Ramsar, Iran, in 1971 (RAMSAR), which involved the conservation of wetlands of international importance, especially as waterfowl habitats.

Additionally, a number of global frameworks were developed to conserve areas of outstanding ecological importance. These involved UN agencies, such as:

- UNESCO (United Nations Educational, Scientific and Cultural Organization), responsible for the biosphere programme
- UNEP (United Nations Environment Programme), responsible for GEMS (Global Environmental Monitoring System)

Private organisations are also involved — for example:

- WCU/IUCN (the World Conservation Union), responsible for designation and categorisation of a range of protected area categories, including World Heritage Sites
- WWF (Worldwide Fund for Nature), a non-government organisation, which works with governments on the management of protected areas

Table 30 is a summary of the aims of conservation and how the various agencies and organisations are involved.

There is a range of ways in which conservation can be approached (Figure 98) — this can be represented as a 'spectrum' of conservation styles and methodologies. Table 31 illustrates some of the changes in conservation that have occurred in the last 30 years.

Figure 98
A balancing act is needed between protection and exploitation

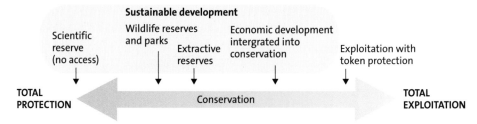

Contemporary Case Studies

Conservation aims

Management category	Conserve and improve hydrological systems	Prevent and control erosion and sedimentation	Conserve and improve timber and related forest resources	Conserve representative sample species (protection)	Habitat conservation	Protect wildlife resources	Conserve genetic resources	Provide opportunities for recreation	Provide opportunities for research, monitoring and education	Improve/perfect environmental quality	Achieve conservation and rural reserve development	Support lifestyles of indigenous people	Promote sustainable rural development	Control exploitation of resources
Biological reserve	▒	▒	☐	█	█	▒	█	☐	▒	▒	▒	▒	░	█
National Park	▒	▒	▒	▒	█	█	▒	█	▒	▒	▒	☐	░	▒
Forest Reserve	▒	█	█	░	░	▒	▒	▒	▒	▒	▒	▒	▒	█
Wildlife/wetlands refuge (RAMSAR)	▒	▒	☐	▒	█	█	▒	▒	▒	▒	☐	☐	☐	▒
World Heritage Site	☐	☐	☐	▒	█	▒	☐	█	▒	▒	▒	░	░	▒
Biosphere Reserve	▒	▒	▒	█	█	▒	█	☐	█	▒	█	█	▒	▒
Managed Resource Protected Area	☐	░	▒	▒	▒	▒	▒	█	▒	█	█	█	█	█

Key: █ = prime aim; ▒ = important objective; ░ = an objective where resources permit; ☐ = not applicable

Table 30 *Conservation aims*

Era	Perception of nature	Relationship with local people	Solutions and technologies	Power relations	Key influences
1960s	Remote wilderness	Local people seen as a threat	Exclusion and protected areas	Alliances with national and government agencies	Colonial conservation; elitist interests
1980s	Ecosystems, biodiversity and eco-regions	People can't be ignored; people to be more of a resource	Buffer zones; a move towards more sustainable-based community conservation	Welfare and biodiversity for people	Sustainable development debate; growing concern for livelihoods
1990s	'Culture in nature'	Align with local people – symbiosis	Alternative protected areas; participatory natural resource management; human rights	Many more alliances at grass roots and bottom-up	Human rights movement and participatory development

Table 31 *Changes in conservation*

The protectionist approach

A protected area is an area of land and/or sea dedicated to the protection and maintenance of biological diversity, and of natural and associated cultural resources. Nearly every country has protected-area legislation and designated sites for protection. In 2002, some 44000 sites met the IUCN definition of a protected area; these sites covered nearly 10% of the land surface of the planet.

Table 32
IUCN management categories of protected areas

Category	Description
I	Strict Nature Reserve/Wilderness Area: protected area managed mainly for science or wilderness protection
Ia	Strict Nature Reserve: protected area managed mainly for science
Ib	Wilderness Area: protected area managed mainly for wilderness protection
II	National Park: protected area managed mainly for ecosystem protection and recreation
III	Natural Monument: protected area managed mainly for conservation of specific natural features
IV	Habitat/Species Management Area: protected area managed mainly for conservation through management intervention
V	Protected Landscape/Seascape: protected area managed mainly for landscape/seascape conservation and recreation
VI	Managed Resource Protected Area: protected area managed mainly for the sustainable use of natural ecosystems

- In Category I, scientific research is of prime importance only when there is 'extreme protection'.
- In Category Ib, biodiversity protection, though important in many protected areas, is not always the primary purpose of a protected area.
- In Category IV, tourism and recreation occurs as a management objective in nearly every category of protected area.
- In Category V, many National Parks have the commemoration of cultural and historic integrity as central elements of management. These areas are often important as tourist destinations.

The concept of a protected area is based not only on the maintenance of environmental services, but also on sustaining natural resources and the maintenance of cultural and traditional attributes.

The growth in the number and extent of protected areas is shown in Figure 99.

Problems with the protectionist approach

Protected areas are the cornerstone of conservation policies in many countries, but they have been criticised for a number of reasons:
- In the poorest countries of the world, there is conflict between economic development and conservation.

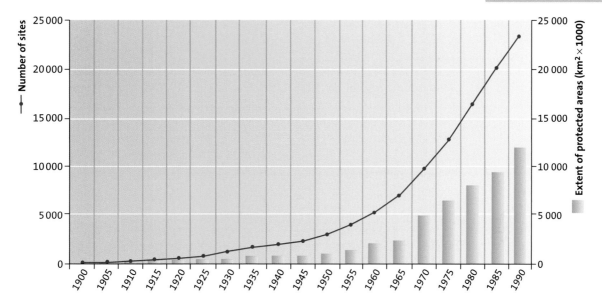

Figure 99 Growth in protected areas, 1900–1990

- Protectionist policies are often narrowly focused and can fail to see that conservation is also influenced by social, economic, cultural and political factors.
- Many schemes are based on the boundaries of countries, whereas ecosystems such as rainforest are restricted by biomes or eco-regions, not political borders.
- Protectionist strategies often rely on the coordination of agencies (both government and non-government), which sometimes fail to work together.

16 Using case studies

Question

Figure 100 gives details about the amount of protected land in the countries of the world. Describe and account for the uneven distribution of the levels of protection.

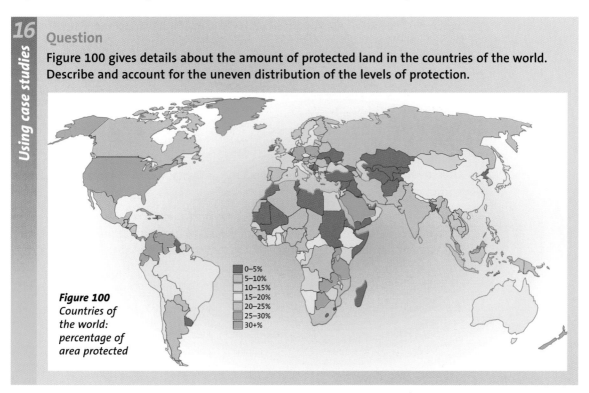

Figure 100 Countries of the world: percentage of area protected

0–5%
5–10%
10–15%
15–20%
20–25%
25–30%
30+%

Guidance

In general, developed nations have larger proportions of protected areas than less developed ones, although some poorer nations (e.g. within Africa) do not fit this pattern. The highest proportion of protected areas are found in North America, Greenland and southern African countries.

In many of the poorer countries, there is a conflict between economic development and conservation. The conservation of large animals, for example, could upset the food supply of a growing local population. Richer countries have to spend larger amounts on protection as land costs are higher. Conservation costs per km^2 in Italy are US$5280, whereas in Brazil they drop to US$61 per km^2.

The future for protected areas

Protected areas may still have a critical role to play in conserving biodiversity, as well as contributing to the priorities of local economic development and poverty reduction in developing countries. However, conservation agencies cannot afford to create more areas without addressing the issues confronting those that already exist — both in terms of their effectiveness in conserving biodiversity, and their negative consequences for residents and neighbouring communities.

Protected areas alone may not be sufficient to tackle biodiversity loss. Strict protection must be seen as a last resort rather than the ideal, i.e. as *one* strand of a bundle of strategies that deliver biodiversity conservation, within the broader context of sustainable development. There is huge untapped potential for increasing the amount of land under conservation — rather than protection. This may well have a greater impact on biodiversity than the creation of new protected areas, and would contribute more to the broader sustainable development agenda.

Managing areas for conservation: sustainable strategies

After 1990, a new model for conservation projects was developed, involving a holistic approach, covering larger geographical areas and combining a global framework with local schemes.

Conservation strategies should be holistic, with strategies involving local economic development as a conservation tool. Examples are the development of ecotourism in Aboriginal reserves of Cape York (Australia), in Ecuador, or the extractive reserves idea in Brazil (collective long-term land use rights given by the government to communities of forest-based people).

With impending concerns about pollution and climate change, larger geographical areas (bio-regions) needed to be conserved, with inter-linking areas across national boundaries and with migration corridors linking these areas. An example is the Yellowstone to Yukon (Y2Y) ecozone in the US (Figure 101). The Y2Y zone stretches an astonishing 3207 km, from Wyoming in the south to the Arctic circle in the north.

It encompasses 1.2 million km². This is an area nine times the size of England. The project links 11 national parks, wilderness areas and ecological reserves along the 'spine' of the North American continent. It is hoped that a coordinated approach will limit development, urban sprawl and the impact of logging and mining.

There is recognition that schemes in LEDCs, where many of the rainforest and coral reef hot spots are found, would require some form of financial pump-priming, i.e. government action to stimulate the economy, from MEDCs (e.g. the Debt for Nature scheme). However, many LEDCs are fearful of MEDC interest in their gene pools, particularly from TNCs.

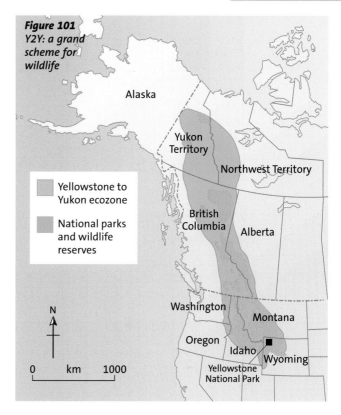

Figure 101
Y2Y: a grand scheme for wildlife

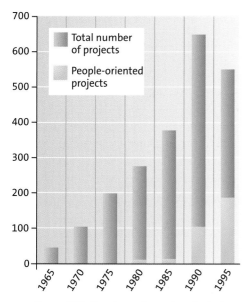

Figure 102 *Numbers of people-oriented conservation projects*

A global 'top-down' framework needs to be combined with bottom-up schemes, involving local people in valuing ecosystems, yet at the same time gaining benefit from them. It has increasingly been realised that simultaneous operations at a variety of scales, from global to local, are required to conserve biodiversity. The number of 'people-oriented' conservation approaches has increased significantly since 1980 (Figure 102).

COMMUNITY CONSERVATION IN KENYA

Case study **17**

'Living with wildlife' in Kathekani

The Kathekani region covers 820 km² and borders the Tsavo National Park East. The present population of the Kathekani region is about 18 000. Agro-pastoralism is the main land use within this semi-arid zone. Bovine trypanosomiasis (spread by the tsetse fly) is the major threat to animal health in the area.

Communities that inhabit the national parks in Africa, live with wildlife that roam in and out of those park areas. The advantages of living with wildlife are often

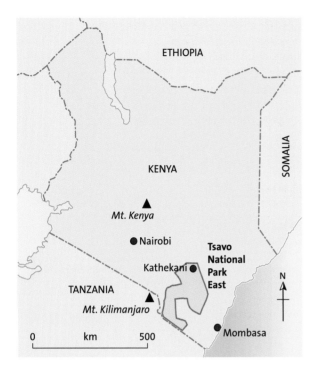

outweighed by the disadvantages. Obvious concerns include trampling of crops and risk of loss of livestock to large predators. Kathekani has an interesting conservation history, not least because of the land-use issues in the area. Technically, the settlers/landowners are squatters on government land.

Changes in conservation strategies for Kathekani

From the 1950s to the 1980s, 'fortress conservation' was the dominant approach. Exclusion policies were combined with enhanced park management. However, there was still a decline in species as the agro-pastoralists (stakeholders) withheld their support, resulting in an increase in poaching.

From the 1980s to 2000, 'community wildlife conservation' was an attempt to link wildlife conservation with sustainable development, using participation and influence from local communities. This more bottom-up approach is based on trade-offs. Local people must gain sufficient access to alternative benefits (e.g. tourism), to offset reduced access to material resources. However, there are some key challenges with this approach:

- economic health is largely based on wildlife tourism — this can be volatile, as levels of tourism are linked to political unrest and perceived risk by visitors
- it leads to cultural disruption in some instances
- there is often an unequal distribution of revenues, combined with limited participation in the schemes

Recent research has shown that educating local communities and encouraging sustainable development have not helped conservation. A team of researchers asked 13 conservation managers, at 16 African wildlife parks, to rate how successfully their protected area prevented wildlife loss.

Successes	Failures
▪ Wildlife conservation projects are changing people's attitudes, in particular some communities are now recognising that wildlife can have an economic value.	▪ Local institutions may lack a democratic structure to implement approaches required by the community plan. The result may mean bias in the decision-making process.
▪ It is estimated that commercial rates per hectare for wildlife viewing are up to four times that for livestock alone. Tourism income in Kenya is growing at around 20% per annum.	▪ There is generally a negative impact on food security and income, especially for the poorest members of a community.
▪ Only limited infrastructure is required in low-tourism areas such as Kathekani.	▪ Some of the improvement strategies do not fit in with local people's priorities. 'I hear they give education bursaries, but where do I get the funds to enable my children to reach secondary school — you see I have no cows to sell' was one complaint.

Contemporary Case Studies

As expected, parks with few people living nearby, good law enforcement and strong public backing were the most successful. However, in other park areas, even when people are taught the value of biodiversity, individual needs (e.g. feeding themselves and their family) will come before any future pay-backs from conservation.

Certain factors will need to change if community wildlife conservation is to succeed. These include the level of economic benefits to both individuals and communities. In particular, a better distribution of income, training, seed groups and rotating credit (where a core of participants agree to make contributions to a fund that is then distributed to the group) is required.

SUSTAINABLE FOREST MANAGEMENT
Case study **18**

A tale of two concepts

In recent years, a have number of forces have led to different approaches to managing forest ecosystems.

- Broadened objectives: foresters are being urged to deal with a much wider range of social and environmental issues than in the past. Management is moving from production objectives to multiple functions objectives.
- Codifying of good practice: policymakers and organisations are developing criteria and indicators to measure forest health and quality.
- Recognition of different interest groups: there is now more understanding that different forests support different stakeholders and require different management systems. There is no single solution to fit all conditions.

Two main approaches have been established:

- The first approach is termed **sustainable forest management (SFM)**, developed from classical forestry and through organisations such as the United Nations Forum on Forests (UNFF). SFM has been developed by forestry professionals, with the primary focus on producing goods and services, with the land under their control.
- The other approach, the **ecosystem approach (EsA)**, has been developed primarily in the framework of the Convention on Biological Diversity. In this case, forest management has been re-orientated to encompass a more collaborative approach.

Table 34
Different approaches to managing forest ecosystems

The precise differences between the approaches are explored in Table 34.

Criteria	Commercial (yield) forestry	Sustainable forest management (SFM)	Ecosystem approach (EsA)
Primary concern	Sustainable commodity production	Balancing conservation with production and use of forest goods and services	Balancing and integrating conservation with the use of biological diversity
Resource management objectives	Based on long-standing government-based traditions; focused on production	Incorporate broader range of environmental and social objectives	A matter of societal choice
Management	One of command and control — 'we manage'	Slightly more open — 'we manage,	Replaced by the concept of 'learning together'
Knowledge	Based on scientific and technological knowledge	Based on expert knowledge, supplemented with broader stakeholder inputs	A more balanced use of scientific and indigenous ideas (local knowledge)

Using the ecosystem approach in India

The evolution of ecosystem approaches in India has caused a major turnaround in the government's stance on the role of communities in forest conservation and management. The following extract from the *National Commission on Agriculture, 1976*, reveals a clear prejudice against local people's forest use:

> Free supply of forest produce to the rural population and their rights and privileges has brought destruction to the forest and so it is necessary to reverse the process. The rural people have not contributed much towards the maintenance or regeneration.

Contrast this with the following statement from the *Indian Forest Plan, 2002–2007*:

> A broader livelihood approach, covering productive capacity, institutional and legal structures, market access and tenure, must be adopted that puts forests into the broader context of rural development. No strategy to conserve the forest ecosystem would be successful, unless the basic needs of the society are met.

There are some possible problems for future forest policy in India:

1 Forest Protection Committees are not statutory bodies and are operating only under administrative order. Their lack of legal standing often leads to legal disputes questioning their rights to protect certain areas.
2 External pressure, in the form of threats from timber smugglers, are accompanied with internal and external community conflicts.
3 The male dominance of the forest protection committees makes them unrepresentative of the needs of the community as a whole.
4 Different communities (and forest departments) merely working together does not highlight the true spirit of the ecosystem approach. Future programmes must involve women and the less powerful sections of the society more fully and efficiently

Hot spots and eco-regions

Hot spots

Biodiversity **hot spots** are areas around the globe where the greatest numbers of species are under imminent threat. These hot spots occupy 2.3% of the land surface (globally), yet are home to 50% of all known vascular plants species and 42% of land vertebrates.

The hot spot approach was proposed by Normal Myers in 1988. Myers suggested that to qualify for hot spot status, an area must contain a large number of endemic species (plants or animals only found in one particular area or region and nowhere else on Earth), and it must be under threat from human activity. There are now 34 hot spot regions worldwide (Figure 104).

Hot spot problems

Some conservationists are now beginning to dispute the value of hot spots. They argue that they do not provide a fine enough resolution, i.e. they are too 'broad-brush' in their approach. Within the Sundaland hot spot (which includes Borneo,

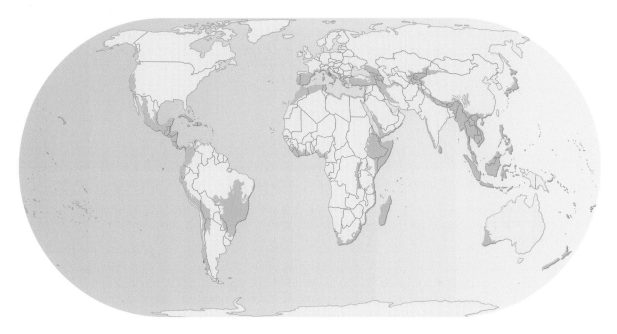

Sumatra and peninsular Malaysia), the WWF recognises 19 smaller and distinct eco-regions, each with a particular diversity of species.

Critics have also identified another problem associated with hot spots — the heavy focus on the *number* of endemic species. The species-rich tropics and subtropics get most attention, while the less diverse, but no less distinctive, temperate and Arctic regions tend to be overlooked.

Introducing the eco-region approach

The eco-region approach is all about conservation on a smaller scale, doing away with the larger and perhaps more unwieldy hot spots. The WWF has identified 867 eco-regions worldwide. This finer-scale analysis can reveal interesting patterns. In an analysis of 140 Indo-Pacific eco-regions published in 2004, those with the highest proportion of endemics turned out to be islands and mountain tops. These relatively isolated settings have meant that new species can evolve. So future approaches to conservation should perhaps adopt a 'highlands and islands' approach — these smaller areas are also easier to protect. The 867 areas were consolidated into 200 eco-regions (the WWF/World Bank Global 200).

Conservation issues

Conservation is complex and raises a number of key issues.

Which areas?

Should we go for a comprehensive representation of each ecosystem type (e.g. the WWF World Bank Global top 200 eco-regions approach) or should conservation be more focused on the hot spots? Other approaches have concentrated on specifics such as endemic bird areas, forest frontiers and endangered animal species (high-profile large mammals). The latest Convention on Biodiversity (183 signatories) focused on endangered plants on the IUCN Red List. There are 16 clear targets for each country to meet by 2010. For the first time, these targets apply to anywhere the endangered plants grow — not only in protected areas.

Which plants and animals?

Do we go for maximum diversity (e.g. the hot spot approach) or for species that may be lower profile but equally important because of their endemic nature or rarity value? High-profile species (whales, tigers etc.) tend to attract more interest and finance than smaller plants and animals, but these may actually be the keystone species.

Design of reserves

There is also the 'SLOSS' debate — single large or several small reserves. Reserves should be large enough to be a minimum dynamic area for an ecological habitat to survive and also support a minimum viable population of species. An equal area divided between several smaller reserves may be less appropriate than a single large one on these grounds, but these reserves may be better buffers against disease, predation or severe natural disturbance. Small reserves are more vulnerable to erosion by human actions/activities. Sprawling linear shapes may cover more diversity of habitats, and allow greater species movement (needed for migration from the possible impacts of global warming). Latest thinking on sustainable management emphasises the importance of a buffer zone, which protects the core conserved area (Figure 105). The core area is protected to conserve biodiversity, the buffer zone is an area of sustainable development, while the transition area is an area of rehabilitation.

Figure 105 *The use of buffer zones around the Great Smoky Mountains Park in the US*

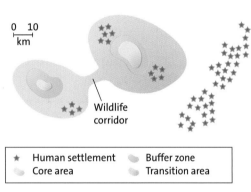

In situ versus *ex situ*

In situ conservation is the maintenance of organisms in their wild state in existing locations, for example national parks and reserves. *In situ* conservation can be difficult when local people are poverty stricken and rely on the local game to survive.

Ex situ conservation usually refers to moving viable populations and managing them in captivity. This is only possible for a small number of plants and animals and is very costly. *Ex situ* conservation may also include gene pools and banks. A successful example is the reintroduction of red kites in Oxfordshire and Gateshead's Derwent Valley.

Legislation and financing

Conservation is a costly business. How should it be organised on a political basis? Should individual countries pay for conservation? If so, this could give rise to tension between MEDCs and LEDCs. What is the role of NGOs (e.g. Greenpeace) in relation to government bodies. There is also the 'top-down' versus 'bottom-up' debate.

Ecosystem futures

There is no doubt that nature is in distress. Although sustainable development has been put forward as a solution to the management of ecosystems, it too comes with a number of challenges:

■ The most biodiverse ecosystems, with high bio-quality, are found in tropical regions, especially in rainforests and coral reefs, where the countries are largely LEDCs.

■ There are conflicts between short-term gain and long-term benefits, as well as local versus distant issues.

■ Many of the products (goods) provided by these ecosystems are not found within MEDCs, which have set up TNCs to exploit them. Short-term profit from goods is at the expense of services and possible future use.

In addition to market liberalisation, globalisation, population growth, increased affluence and urbanisation, ecosystems may be facing a greater and even more damaging threat — climate change. Table 35 paints a bleak picture for the outlook of the world's ecosystems.

Farming in the UK, for instance, is undergoing significant changes since the introduction of the Entry Level Stewardship (ELS) scheme developed by DEFRA (Department for Environment, Food and Rural Affairs) in 2005. Under the ELS, farmers are paid for carrying out environmental improvements to their land, with the goal of restoring wildlife. Importantly, the scheme promises to link-up fragmented habitats, so this should encourage improved ecosystems and wildlife stability. A range of regeneration strategies are being developed, including hedgerow management, wildflower strips, buffer zones around ditches and ponds, field corner management, seed strips, conservation headlands and beetle banks.

Table 35 Relating temperature increase to food production, water resources and economic damage

Temperature increase (°C)	Food production	Water resources	Economic impact
0.7–1.0	Developing countries suffer small but significant falls in crop yields	Between 400 million and 800 million people have an increased risk of water shortages	Net losses in GDP for a number of developing countries
1.0–2.0	Number of people at risk of hunger increases significantly	An extra 1.5 billion people at risk of water shortages	GDP losses worsen; damage in regions such as Africa becomes significant
2.0–3.0	Big rise in populations exposed to hunger — as many as 5.5 billion people	Up to 3.1 billion people at risk of water shortage	Losses begin in global GDP
3+	Entire regions become unsuitable for producing food	Water stress becomes even worse for billions of people	Global damage to GDP increases substantially

Conservation and ecosystem management are gaining global political currency, but successful conservation must spring from a partnership between the government and the people.

Such affiliations should draw from local creativity and local knowledge, and needs to encourage stewardship across entire ecosystems. This must be underpinned by resources from the public sector, particularly for beleaguered countries who have long been victims of exploitation. Protecting biodiversity — whether in remote forests or in the concentrated hot spots on land and in the oceans — is achievable. Many of the necessary measures are affordable and should supply local economic benefits. Whether we deliver on this is up to our generation. By the time the next generation has a chance to decide, it may be too late.

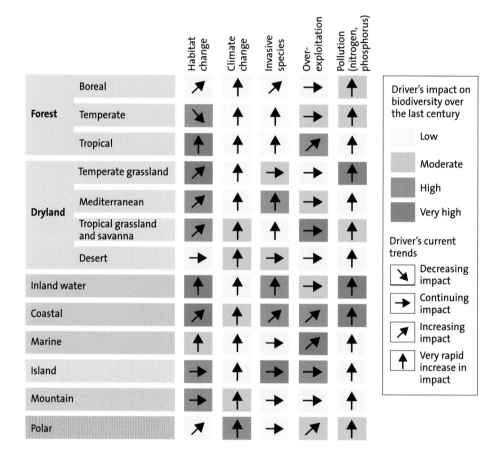

Figure 106 Main causes ('drivers') of change in biodiversity and ecosystems

Summary

1 Biodiversity is immensely important, yet it is a complex idea to grasp. Unfortunately, it cannot be measured with simple indices or reduced to a single number.
2 Statistics used to measure biodiversity losses can be variable. In many instances, it is not possible to just accept the figures given, since they are reported in a confusing or contradictory manner. Always check the data and their source — quote with care.
3 Biodiversity has tangible value in terms of goods and services derived; the financial (direct use) values can be more easily estimated, than the non-use or intangible elements.

Contemporary Case Studies

4 Biodiversity varies from place to place and on different scales: locally, regionally, nationally and internationally. Generally, the areas of the world with the highest levels of biological productivity (land or sea) have greatest diversity.

5 A significant challenge for conservationists is to reduce rates of biodiversity loss (or to treat the main causes: habitat loss, invasive alien species, over-exploitation, pollution and climate change).

6 There is a complex arrangement of global frameworks to manage and conserve ecologically important areas. Examples include hot spots, ecoregions, national parks, RAMSAR sites, biosphere reserves and forest reserves.

7 The 'protectionism' approach (especially favoured in the 1980s) is now viewed by many as a last resort, rather than the best current strategy. Twenty-first-century conservation requires a portfolio of approaches, including more locally based 'community' conservation, sustainable management and integrated conservation within a framework of economic development.

8 Loss of habitats and ecosystem quality is *not always* a function of wilful exploitation of environments by individuals, organisations or governments. The underlying *root* causes are more often to do with poverty, improving quality of life, development, conflict over resources and shortage of space.

Websites

http://www.gbif.org/
The Global Biodiversity Information Gateway.

http://www.undp.org/biodiversity/biodiversitycd/bioImport.htm
The UNDP have some good information about global biodiversity issues.

http://www.nhm.ac.uk/research-curation/projects/worldmap/
The Natural History Museum offers some interesting perspectives on biodiversity.

http://www.ukbap.org.uk/
UK Biodiversity Action Plan website.

http://darwin.eeb.uconn.edu/eeb310/lecture-notes/reserves/node8.html
Brief outline of the SLOSS debate.

Examination advice

The case studies featured in this book can be used for a variety of questions across a range of examination boards and specifications.

Handling the 'standard' ecosystems questions

Describe the distribution of ecosystem X

In this instance, 'X' could be, for example tropical rainforests, coral reefs, tropical grasslands.

This question will inevitably require you to use a map from which you can describe the distribution. Remember, 'distribution' refers to the arrangement of something — in this example, an ecosystem type. Use the resources effectively and state precise locations, quoting latitude and longitude where possible.

Some exam questions might show a map with a number of 'ecoclines', i.e. transitions between different types of ecosystem. An example would be the transition from tropical rainforest to seasonal rainforest.

Discuss the relative importance of physical and human factors in controlling the distribution of ecosystem X

The primary controls will all be climatic (controlling distribution at a macro scale). In particular, consider temperature, rainfall and length of growing season. There may be other physical controls or limiting factors that operate more locally. These could include microclimate (aspect, relative position on slopes, altitude etc.), soil type (quality, depth, drainage and nutrient status) and competition from other vegetation.

Human factors will be important, but only at a local level. Factors might include eutrophication, grazing pressure, urbanisation, management regimes or deforestation. Although only localised, these factors may have a significant impact. For example, deforestation in tropical rainforests leaves soils stripped of nutrients and prone to erosion.

Examine the factors influencing variations in ecosystem productivity

Your response to this question requires a definition and clear understanding of the idea of productivity, and how this can vary on a global scale. It is important to know a rank order of ecosystem productivity (i.e. reefs, tropical rainforest, marsh, seasonal forest…desert/ice/rock). Refer to table 2 on page 8. The main controls on productivity will be climatic, although people will also influence productivity at the micro-scale. You could also show that productivity can vary temporally, as well as spatially. Examples should be drawn from a range of ecosystems and at different scales.

Handling the mini-essay (8–10 marks)

Here are some typical examples.

(1) With reference to one or more ecosystems, describe and explain how human activity might influence energy flows

Your answer might use deforestation of a forest ecosystem, when energy converters are removed from the ecosystem. Deforestation affects the nutrient cycle, in that there is less organic matter to be recycled and so less humus and nutrients are added to the soil. Deforestation might also affect the hydrology of an area. You could also use a marine ecosystem, for example coral reefs.

The best answers will show a confident use of geographical language.

(2) Outline the biotic and abiotic elements of a small-scale ecosystem you have studied

- ▪ 'Biotic' deals with the living parts of the ecosystems (i.e. all organisms).
- ▪ 'Abiotic' factors are geological, geographical, hydrological and climatological variables, for example soil, geology, temperature, hours of sunlight, salinity and pH.
- ▪ 'Small-scale' ecosystems could refer to a pond, woodland or perhaps a sand dune.

Your answer should include a description of the main biotic and abiotic components and a realisation that these are difficult to separate, since an ecosystem is a place where all of these variables meet and interact. Changes in one biotic or abiotic variable can upset the whole ecosystem balance.

(3) Referring to an example, outline the social and economic impacts of the deforestation of a tropical rainforest

See Table 36.

A strong answer will consider social and economic impacts in a balanced framework and will use specific examples (not just 'the Amazon') to support the response, including detailed facts and figures.

Table 36 Impacts of deforestating a tropical rainforest

Social impacts	Economic impacts
■ Loss of traditional lifestyle for indigenous inhabitants/loss of land	■ In many countries with tropical rain-forests, there is rapid population growth and a need for employment, e.g. in Indonesia, logging employs 700 000 people directly
■ Threat of disease from in-comers to indigenous population of the forest	
■ Migrants moving into the forest to farm often face a very low standard of living	■ Wood and wood products produce export earnings and support other industries, such as furniture manufacture; hardwoods such as mahogany and teak are highly valued
■ Pollution, as trees are burned, affects the health of the population	
■ Some locals have been killed trying to defend their land from development	■ Export earnings help to pay off large international debts

Drawing key diagrams

There are a range of important diagrams that you should be able to use in an exam question. Not only will the diagrams have to be reproduced accurately and with appropriate labels/annotations, but also constructed quickly under timed conditions. Figure 107 shows some examples.

Figure 107 You should be able to draw these diagrams quickly, from memory, under timed conditions

Basic soil profile diagram

Nutrient cycling (Gershmehl)

Basic succession model

EXAMPLES OF DIAGRAMS YOU SHOULD KNOW

Conservation spectrum

Total protection

Total exploitation

Trophic levels/pyramids of biomass

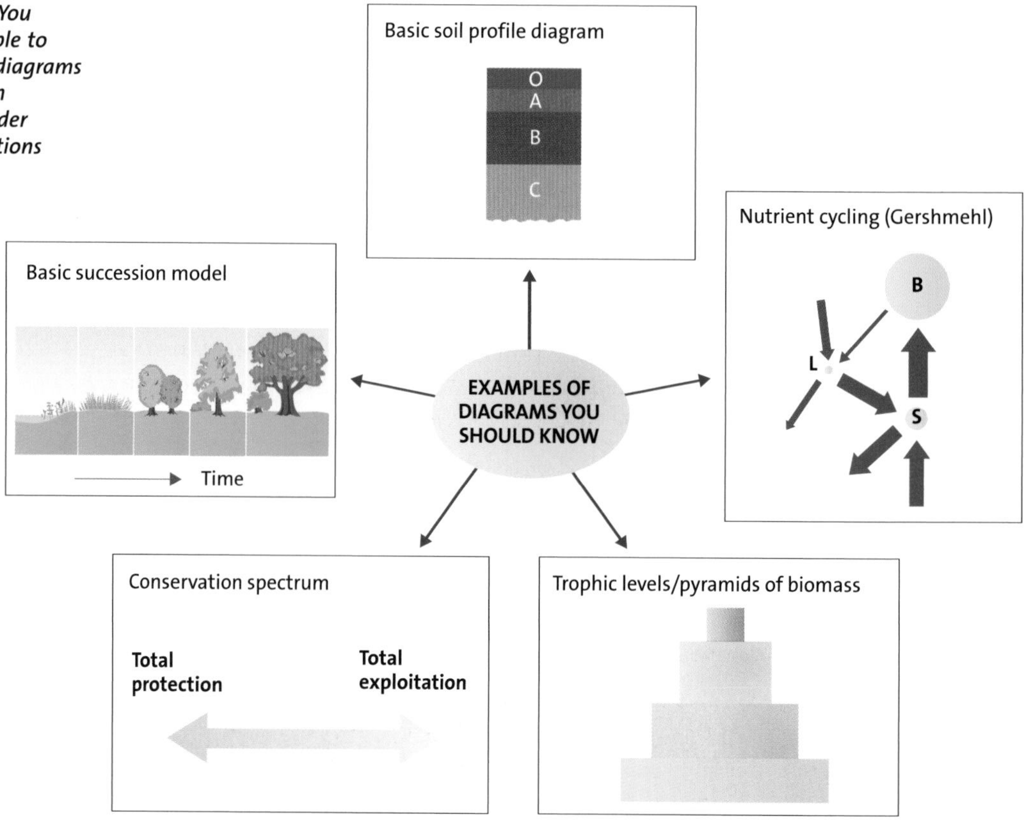

Handling resources (including synthesis and analysis skills)

Another essential basis for exam preparation is being able to handle resources. The images in Figure 108 are from the *Atlas of Our Changing Environment* (*UNEP, 2005*) and show a section of Amazon rainforest.

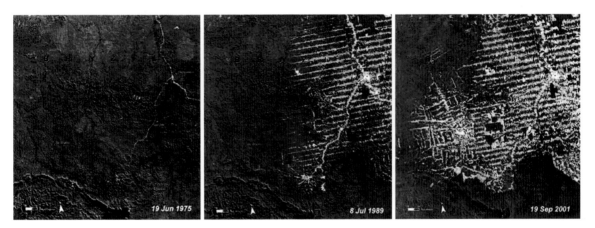

Typically, an exam question would use these resources to prompt a lead-in question with a relatively basic command word. An example could be: 'Describe the changes shown in the photographs from 1975–2001'. You would need to comment on the areas of development in the east of the image following the course of the river (1989–2001) Also, you would need to note how the rate of change (i.e. deforestation) has accelerated. It appears that natural woodland has been replaced by more intensive agricultural production.

If resources include a scale, then make use of this to develop your answers further, with specific numbers to back up your comments.

Logical follow-on questions include: 'Account for the changes' and 'Suggest ways in which the rate of change could be slowed or even reversed'. These use more sophisticated command words, requiring greater levels of knowledge, understanding and application.

An alternative resource may be an extract of prose or writing. This will usually be around 100–200 words. The source of such a resource may be a newspaper, a magazine or a report from the Internet. For example:

> Conservationists tend to be conservative. But not always. Here is one strikingly non-conservative conservation idea.
>
> People and wildlife don't get on too well together. Large mammals, in particular, have a hard time at the hands of humanity. Their habitat gets taken for farms, their bodies for dinner and their heads for trophies. As human populations grow, the pressure increases, and it seems to decline

Figure 108
The impacts of deforestation, Rondonia, Brazil, 1975-2001. Note the development of the 'fishbone' pattern (1989 and 2001) indicating an increasingly agricultural land use.

■ There is a conflict between people and wildlife

- Economic wealth means that conservation can be given a higher priority

- There is a contrast between LEDCs and MEDCs in terms of population (growth and movement), as well as affluence

- North America once had large mammals, but these have been lost in the last 13 000 years

- Reintroduction of once native animals to North America may be a viable method of conservation

only when people are rich enough to focus on the aesthetic, as well as the economic, possibilities of wild beasts. Often, such aesthetic appreciation thrives best in the safety of the city, rather than in the rawness of the wilderness.

Observing all this, a group of conservation biologists, led by Josh Donlan of Cornell University, have made a modest proposal in this week's *Nature*. They suggest a piece of ecological arbitrage.

Africa and Asia are continents where wildlife is under particular pressure. Their human populations are growing and their people are not yet prosperous enough to make conservation a higher priority than simply getting by in life. But many of the world's endangered mammals live in Africa and Asia. In North America, by contrast, rural populations are shrinking, people are rich enough to care about wildlife, and many of them do. Moreover, most of the large North American mammals that existed when humanity arrived in the continent are now extinct. When the first immigrants entered North America at the end of the Pleistocene epoch, more than 13 000 years ago (how much more, is the subject of vigorous debate), they found a continent full of large mammals — elephants, lions, cheetahs, camels, horses and more. Within a few thousand years, most of these animals were gone, probably the victims of overhunting. Their ecological niches are therefore wide open for occupation. What could be more logical, Dr Donlan suggests, than introducing endangered Old World mammals into the New World, thus saving them from extinction, while returning wild America to something like the state it was in before *Homo sapiens* took up residence?

The Economist, 20th August 2005

The skill of synthesis is all about understanding and summary. Start by deconstructing the piece and making about five or so short bullets points, as shown in the example above. Once this is completed, you may want to develop the ideas further into a concept diagram (Figure 109). This again helps understanding.

Figure 109
Concept diagram

	USA	Africa
Options/ opportunities	• Affluent economy can make conservation more of a priority (even 'pay' for animals)	• Relocate existing populations of large animals (once native to North America) — is a conservation strategy
Realities/ challenges	• Rural depopulation (rural urban migration) • Large mammal loss and extinction in historical times	• Rapid population growth = conflicts over space and resources • Developing economy which is resource hungry • Conservation given a low status

These techniques can be used to handle resources within the exam, but can also be used for pre-release issue analysis style materials.

The longer essay answer

Standard advice on essay planning should be followed, namely that the essay should have a clearly identifiable, three-part structure.

1 A brief introduction: this may offer a context and appropriate definitions.
2 The expansion: this is where the argument or points are developed in a sequence of logical paragraphs that incorporate supporting examples and case study materials.
3 The conclusion: keep this short (one paragraph) and offer a concise, but effective, summary to the main essay, referring back to the original question.

An example question

Referring to two contrasting global biomes, identify the links between the climate and vegetation

Introduction

A sensible start might be to give an overview of the term 'biome', and explain briefly the distribution of important biomes. Then justify the choice of two contrasting types; in this case, we have gone for tropical rainforest and tundra. These are not only contrasting in terms of geographical location, but also in terms of productivity, communities.

Expansion

Tropical rainforest (TRF)

- TRF grows in humid tropics, where rainfall often exceeds 2000 mm and is evenly distributed throughout the year (although there might be a short dry season).
- Temperatures are uniformly high, with a mean monthly temperature of 27°C. Temperatures are suitable for continuous vegetation growth.
- Trees have slender trunks and thin, smooth barks to allow maximum water loss and because there is no need for protection from frosts.
- Leaves often have drip tips to enable them to shed water and encourage transpiration.
- Buttress roots allow massive uptake of water.
- Up to five layers of vegetation are present.
- Undergrowth is sparse, except along river banks and in clearings.
- Leaves are grown and cast off continually, to reflect the year-long growing season.
- Climate encourages great diversity of species — often more than 200 different species per hectare.

Tundra

- Tundra grows in high latitudes, where temperatures are below freezing for at least 6 months of the year.
- Precipitation is low — around 250 mm per annum — falling mostly as snow.
- Land is subject to permafrost, with some surface thaw in the summer.
- Winds are strong throughout the year.
- Vegetation is sparse, generally with no trees.
- Species diversity is low.

- Most plants are mosses, lichens, grasses, sedges and dwarf shrubs.
- Only one layer of vegetation is evident.
- Most species are frost- and wind-resistant.
- Plants have shallow roots to cope with permafrost.
- Plants are adapted to conserve moisture; they have thin and often hairy leaves, with thick cuticles.

Conclusion

The question demands that the essay looks at the links between climate and vegetation. For the examples above, the links are very strong, with the plants adapted to the particular climatic regime.

Index